QUALITY MANAGEMENT PERSPECTIVE & APPROACH

Managing and improving quality in China, and elsewhere in the world

CHRISTOPHER LOURENS

EMPEROR

First published in Great Britain by
Emperor Books
Second House, Brookfield Close, Redditch, Worcestershire B97 5LL
Email: emperor.books@outlook.com

British Library of Cataloguing-in-Publication Data
A catalogue entry for this book is available from the British Library.

Hardback ISBN 978-1-9996358-0-0
Paperback ISBN 978-1-9996358-1-7

CONTENTS

PART 3: COST & RISK

6. QUALITY COSTS & ASSOCIATED ACTIVITIES

7. PRACTICES HAVING MAJOR BEARING ON COST & RISK

PART 4: ENABLING, SUPPORTING & IMPROVING PERFORMANCE

8. OPERATIONAL EFFECTIVENESS

9. SUPPLIER QUALITY ASSURANCE

PART 5: VARIABILITY, CAPABILITY & QUALITY TOOLS

10. PROCESS AND PRODUCT VARIABILITY & CAPABILITY

11. BASIC QUALITY TOOLS

ANSWERS TO QUESTIONS

FOREWORD

Good management drives success in most organisations, whether they are industrially based or services based. The priorities of managers vary through time. Sometimes the need is simply to get the organisation into being; then to produce a product or service; then to produce it at a competitive cost. In market economies the cost and prices need to be competitive. In all societies quality eventually emerges as the differentiating factor:

Is the product or service efficiently produced?

How well does it meet its need?

Are the processes best suited to its production?

Are the necessary standards being met at each stage?

Are the processes self-correcting when defects are found?

As firms increase in size, how does management keep itself informed that the organisation is fully efficient and what improvements need to be made?

Those and similar questions are answered by good Quality Management in all firms, wherever they are and in whatever society they are placed.

The author, Christopher Lourens, has extensive experience in Quality Management and has worked in many significant firms, in both the West and in China. In recent years he has been committed to helping Chinese organisations improve their Quality Management, and in doing so he has obtained insights into how many such firms can improve their operations by means of enhanced Quality Management. This book sets out the results of his experience. It is written from the perspective of an expert in his field, who has a wide knowledge of his subject, and who feels that he has something more to contribute to many organisations in China and for that matter, elsewhere as well. His style is direct and honest, leaving it to others to work out how best to use his knowledge and insights.

This excellent book is ideally used by all who seek to understand Quality Management and improve the quality of their product or service. Such organisations might be industrial firms producing a product, services based such as universities, or even administrations. The principles of continuous improvement are applicable to all. However, it is especially suited to management executives and management students in China or those involved with companies in China. Ideally it could form part of management studies and for groups which can discuss how best to apply its findings and suggestions to improve their organisations.

All managers worldwide can learn from one another. This author has generously made it possible for Chinese managers especially, to do so if they wish, by drawing on this unique knowledge to measure their current standards against best practice, so as to identify possible improvements in their own organisation.

Strongly recommended!

Emeritus Professor Roy Lourens AM, PhD, FAIM
Former University Vice Chancellor, President of the Australian Institute of
Management (WA)

PREFACE

Much of the material in this book could be used by a wide spectrum of those wanting to understand Quality Management and to improve the quality of an organisation's processes, products and services, and, while many of the issues affecting quality management in China are unique to China, some are not so unique and apply elsewhere as well. Readers can reflect on this and take note as necessary.

The culturally rich world of China was isolated from the development of modern quality management until a new economic development strategy was launched in late 1978. The adoption of the "Reform and Open-door Policy" in the People's Republic of China gave entrepreneurs permission to start businesses, and the country began opening up to foreign investment, foreign management knowhow, and Quality Management.

In 1980 the Chinese government started to promote the importance of Total Quality Management in industrial economic development. In subsequent years, further emphasis was placed on the management of quality, notably the "Product Quality Law of the People's Republic of China", holding manufacturers responsible for the quality of their products, and the adoption of foreign advanced standards including ISO 9001, the International Standard for Quality Management Systems, to which companies could achieve quality management certification (published in China as GB/T 19001), and that is highly valued today. In fairly recent times in the relatively short history of modern quality management in China there has been national encouragement in performance excellence through the "China Quality Award".

Since the 1978 Reform, economic progress has been rapid yet experience within Chinese industrial manufacturing organisations often reveals that present implementations of quality management generally indicate a non-realisation of its full benefits. It is not uncommon to find this in both state-owned and privately-owned mainland-based industrial organisations. There are exceptions seen in technology transfers where quality procedures and process controls have been transferred, however, these means of quality management are likely to be influenced by deeply embedded ways and affected by a multitude of daily problems. They may be subjected to quick fixes and walk-arounds, and suffer a high probability of gradually lapsing into a less effective state.

There are many things that need to be addressed; prominent is the attitude and behaviour towards quality. Organisational culture influences the attitude and behaviour of company leaders and that of the employees. It affects motivation, human performance, and how new ideas and continuous improvements are encouraged. It shapes how people within the organisation interact with each other and the actions that they take. Unless the values of a quality management culture become embedded in the organisation's culture, the true intents and functioning of modern quality management will be hindered.

Furthermore, each organisation is unique in terms of management practices and processes that are employed to create and deliver its products and services; the leadership for developing a quality management implementation solution that is effective for the organisation resides with its management, and the operation involves everyone in the organisation, including its management. The quality management solution will contain activities that strive to eliminate and prevent the cost and risk of poor quality, and activities geared towards improving quality and process performance – these may well include elements from improvement related initiatives such as found in Lean principles, Kaizen, ISO certification, Six Sigma and supply chain management.

There is a lot that the organisation must learn and understand before the full benefits of Quality Management can be realised and sustained, and the leadership and participation of management in the various activities to improve product quality and achieve better quality management is most necessary. This is because managing quality and its improvement involves major decisions and changes that impact on resources, and, most significantly there is a positive correlation between the managers' degree of demonstrable interest in quality and the attention that employees pay to quality.

Quality Management Perspective & Approach is an easy-to-understand sourcebook for evolved and improved quality management – a resource for all who wish to learn what it takes to improve an organisation's quality-affecting processes and the quality of products and services. It presents an insight into matters that affect quality management within industrial manufacturing organisations in China.

Aspects that make quality management an integrated part of an organisation's overall management system are identified, important values of a quality management culture are described, and relevant successfully established "best practices" and frequently employed quality tools and practical techniques are explained. Observations are expressed, and narratives are included of experiences that illustrate various approaches that have been employed to overcome quality deficiencies, improve product quality and enhance quality management.

A direct style is employed – the manner of expression is intended to serve to make clear the issues that are considered to be detracting from the advantageous usage of improved quality management. The express wish is to advance transformation to a beneficially effective quality management.

At the end of each chapter are questions for the reader to reflect on and attempt answering. Answers to these, provided at the rear of this book, are aimed at complementing the contents of each chapter to enhance understanding of the many aspects of Quality Management.

It is hoped that the reader will benefit from obtaining an understanding of practices, tools and techniques and also reach a good comprehension of the intent behind them.

Christopher Lourens, MMXVIII

1. INTRODUCTION

1.1 Shared understandings, characteristics and determinations

The standard of quality management in industrial manufacturing organisations in the West varies a great deal; some organisations have a very high standard and others do not. There is much to learn from organisations that have persevered and developed successful quality management practices of very high standards that really bring results, and it is always instructive to compare our own efforts against best practices elsewhere, to see where and how we might improve our own efficiency and effectiveness.

We need to look at those organisations across countries and cultures in which quality management truly brings results. In these organisations the nature of implementation of quality management may differ according to their business scope, goals, and needs, but they share certain understandings, characteristics and determinations; these can be summed up as follows:

- Quality (see Appendix 1 for the definition) is of key importance to a successful business; poor understanding and attention to customers' expectations concerning the entire contact and relationship with the business places risk on the long-term survival of the business.
- The key purposes of the quality management system are understood to provide confidence in the ability to meet customer requirements, to serve as a quality problem preventive tool, and to provide a cost or differential based business advantage.
- Quality management is fully supported by top management; top management's commitment is evident in their everyday actions.
- The pro-quality actions of managers and supervisors, the empowerment of employees and how they are recognised and rewarded, are essential in sustaining quality-orientated behaviour.
- "Values and norms" are espoused and practised throughout the organisation that constantly influences quality-orientated behaviour.
- Processes are understood to the extent that their performance can be optimised.

- Implementation of effective quality management is accelerated, and gains are most successfully held when there are senior-level "quality champions" in the organisation – better still is when there are "quality champions" present at every level. These individuals drive the quality movement, and have the ability and tenacity to overcome those many obstacles that sap energy from positive momentum.
- Facts, evidences and accurate data analysis are used for decision making.
- Relationships with suppliers and sub-contractors require constant management to optimise quality, cost and delivery performance.
- The management of quality and quality improvement requires constant work and the involvement and continual education of all employees. Organisations that rely on past achievements and make no effort to improve their status slowly fade away and eventually die.

The successful industrial manufacturing organisations in the West follow the principle that every management practice and process must produce a result that in some way enhances the business. This thinking has become more and more emphatic with the increase of national and international competition; today's customers demand more than ever before, and customers will switch for better price and/or better service and/or better product quality. And customers do not view product quality simply but rather as a complex arrangement of exclusive dimensions, i.e., conformance, performance, features, reliability, durability, serviceability, aesthetics, and perceived quality.

The issue on the minds of company managers is that, if their company cannot distinguish itself as a supplier of choice in its market, its customer base will erode and their company will find it more and more difficult to attract new business and to prosper in the long term.

In the West, market/customer requirement and expectation initially caused many organisations to pay serious attention to implementing quality management systems and, since the 1980s, ISO 9001, the International Standard for Quality Management System Requirements, has played a central role in the implementation of quality management systems in organisations throughout the world.

Over time, the initial driving force of market/customer requirement and expectation has been joined by the driving forces of quality management benefits and advantages. These benefits and advantages come by way of the sensible and systematic control and prevention orientation practices and processes of quality management, and its focus on continuous improvement.

Business managers ask themselves questions such as:

- What are the key practices which enable our business to outperform its competitors?
- What drives the sales figures?
- What drives the costs and the cash flow?

They see that quality management is built into each of these by way of, to cite some examples:

- The steps taken to ensure and enhance customer satisfaction.
- Responsiveness to customers' requests, questions and concerns.
- The intention to ensure that the product or service performs right first time.
- Actions taken to reduce costs (prevention of rejects, scrap and waste of time and effort).
- Actions taken to ensure on time delivery of the product or service.

The better industrial manufacturing organisations in the West now accept quality management practices and processes as an effective and essential part of the management practices and processes that are established to enhance the business and bring business success. In other words they are regarded as essential management processes that bring advantages and benefits that lead to improved profitability and competitiveness, improved customer satisfaction, and improved operational effectiveness.

There are organisations in China whose top management realise that there are benefits to be reaped by embracing quality management and they are motivated to develop quality management. Their motivation is also stimulated from the increasing threat of competition from locally manufactured products, and the threat of the superiority of the quality of some foreign products.

1.2 Factors influencing quality management progress in China

Modern quality management in China has been largely driven by government initiatives. Key initiatives since 1980 are listed in table 1-1.

Table 1-1: PRC government mandated key quality management initiatives

March 1980, the State Economic Commission issues Provisional Regulations for Total Quality Management (TQM) in Industrial Organisations. These regulations emphasize the important role of TQM in industrial economic development and stipulate its method of implementation.
December 1988, China adopts the ISO 9000 series of standards which includes ISO 9001, the International Standard for Quality Management System Requirements, published in China as GB/T 19000 and GB/T 19001.
August 1992, the State Council issues "Decision on Further Strengthening Quality Management". This emphasizes the meaning of quality management.
February 1993, the National People's Congress adopts the "Product Quality Law of the People's Republic of China" which holds manufacturers responsible for the quality of their products.
December 1993, the State Economic and Trade Commission, the State Science and Technology Commission, and the State Technical Supervision Bureau jointly issues "Regulations of Adoption of International Standards and Foreign Advanced Standards."
1998, the China National Accreditation Council for Registrars signs the Multilateral Recognition Arrangement (MLA) for international acceptance of ISO 9001 certificates of the International Accreditation Forum (IAF-MLA) on trade in manufactured goods with 16 countries.
July 2000, the National People's Congress amends the "Product Quality Law of the People's Republic of China". The amendment is aimed at strengthening quality supervision and control, raising product quality level, clarifying product quality liability, and protecting the rights of consumers.
The Product Quality Law requires that manufacturers and sellers set up self-supervising quality control systems, and strictly enforce the rule that quality control be assigned to every position in the course of manufacturing and marketing; this provision strives to ensure product quality through the mandatory application of quality control systems in enterprises.

April 2001, the State Council consolidates focus on quality matters (i.e., quality management, metrology, standardisation, certification, accreditation, and administrative law enforcement) and forms a ministerial-level body, the General Administration of Quality Supervision, Inspection and Quarantine (AQSIQ). Under the AQSIQ the State Council establishes the Standardisation Administration of China (SAC) and the Certification and Accreditation Administration (CNCA).

- The SAC supervises and co-ordinates standardisation across China and represents the PRC in international standardisation organisations including the International Organisation for Standardisation (ISO) and the International Electrotechnical Commission (IEC).

- The CNCA is responsible for the administration of the China Compulsory Certification and Safety License System. Activities includes drafting laws, regulations and rules related to certification and accreditation, safety license, hygiene registration and conformity assessment, and the co-ordination and guidance in their implementation. CNCA represents the PRC in the field of conformity assessment in relevant international bodies.

2001, AQSIQ and SAC commissions the China Association of Quality (CAQ) to develop a model of performance excellence to promote quality and business excellence. This gives rise to the launch of the "National Quality Management Award".

August 2004, GB/T 19580-2004, the National Standard of Criteria for Performance Excellence is published.

2006, the CAQ renames the "National Quality Management Award", the "China Quality Award".

March 2012, GB/T 19580-2012 supersedes GB/T 19580-2004; GB/Z 19579-2012, Guidelines for the Criteria of Performance Excellence, is issued.

December 2016, in concurrence with the improved updated ISO 9001:2015, PRC's direct equivalent is released – GB/T 19001:2016, Quality Management Systems Requirements. This replaces GB/T 19001:2008.

It can be seen from the key initiatives that quality management has steadily grown in importance in China. Acquiring accredited certification to GB/T 19001, the quality management system standard, is valued by enterprises. And a growing number of top management have aspirations for their companies to achieve the highly prestigious "China Quality Award". Many in top management have attained an awareness of the benefits attributed to KAIZEN, Lean Manufacturing and Six Sigma – initiatives explained in Appendix 1 – and they are enthused by the idea of "zero defects".

Nowadays, practically all prominent industrial manufacturing organisations in China carry certification to GB/T 19001; however, implementations of quality management frequently indicate a non-realisation of its full benefits. Close scrutiny often reveals that the fashion of implementation of some of the essential processes of quality management suggests that the intent behind them has not been fully understood.

Maybe it is not clearly evident that;
- effective quality management employs risk-based thinking (by way of contract review, design review, and management review);
- a key purpose of quality management is quality improvement (by way of determining, deploying and working towards quality objectives, and by monitoring, measuring and using information not only from production but from a multitude of sources, including quality audits, for corrective and improvement action);
- quality management concerns quality problem prevention (by way of quality planning, design review, quality awareness education, and preventive action);
- the processes of quality management are all geared towards operational effectiveness.

For some organisations in China, it is seemingly more important to create good impression by claiming "compliance" with GB/T 19001 (ISO 9001), and being able to show off their GB/T 19001 certificate, than actually working at obtaining operational benefit and advantage from effective implementation of quality management processes.

Generally, the continuous improvement engine that should be at the heart of quality management can be improved.

There are Chinese companies in which top management has good intention to achieve excellence in quality. Top management pass this intention to middle management and genuinely express and stress their hope for success, yet, even within these organisations, the outcome can be disappointing when the quality management processes are not practised in the routine ways of working.

A few middle managers may accept and commit to the intentions of top management while a large number accept in a compliant and even an unenthusiastic manner, and others, affected by preconceived notions, may deride the idea. From this mixed pool the idea is then passed on to first-line supervisors as an instruction. It is not surprising that the employees, the workers who are most effected by the idea, who are expected to grin and bear most of the change and work-effort, tend to avoid involvement and even ignore what they may perceive as having no benefit to them or yet another pointless scheme. Unfortunately, management follow-though tends to be weak or not forthcoming.

The lack of management commitment to and involvement in the job of making quality management a success is a major stumbling block in the advancement of quality improvement. Failure to educate managers with regards to operational benefit and advantage of quality management would influence their commitment, and this lack of commitment would result in unsatisfactory management involvement.

What about the influences of Chinese culture? The rich Chinese cultural values and beliefs have given rise to Chinese traditions that shape how people interact with each other (respect for the senior and the aged and the hierarchy of authority); they preserve personal identity, a sense of self-worth, and work in favour of building good relationships.

The quality management principle of engaging and involving employees necessitates "managing them effectively by empowering them". Whereas this may be likely to happen within the higher levels of management in China, it is not likely to happen at the level of the ordinary employee. The polite submission and respect to those in power and authority positions discourages this type of participatory quality problem management and quality improvement philosophy that forms the basis of modern quality management. Also, continuous improvement necessitates pointing out problems and making changes for the better, but face-saving and hierarchical position creates reluctance, and there is a preference to follow the peer group and to maintain the status quo.

Another factor is that, unlike the industrial workforce in the West that have been exposed to many generations of industrialization, many of the present generation of industrial workers in China come from poor rural farming communities. To them, modern quality management concepts have little real meaning. Furthermore, the centuries-old custom of long-term planning has instilled in many a cautious approach to new concepts; this does not work in favour of modern quality management and the need to use feedback and accept change to improve process, quality and customer satisfaction.

The influences of Chinese culture and centuries-old customs, the fact that some managers are resistant to accepting certain quality management principles, and that many employees do not understand quality management, would impede or slow progress in the successful implementation of quality management. However, the collective and group oriented aspects in Chinese culture could help the development of quality management. It is considered that if shared beliefs in the value of quality and team participation are developed in the workplace, peer group pressure could ensure that the quality movement is strengthened; intensive and continual training in modern quality management concepts and practices, and their adaption for practical implementation would help with their acceptance and effective use.

1.3 Integral parts of managing and improving quality

Our vessel, Excellence Evolution, figure 1-1, is equipped for the continuous quality improvement journey. The journey begins when top management engages one or more of the core drivers with full commitment. Such core drivers could arise from recognizing the need to improve quality due to market competition, or the requirement or expectation from customers that the organisation has ISO 9001 (GB/T 19001) quality management certification, or the realisation that there are huge economic and operational benefits to be gained through the reduction of loss caused by sub-optimal quality processes and elimination of poor quality products and services.

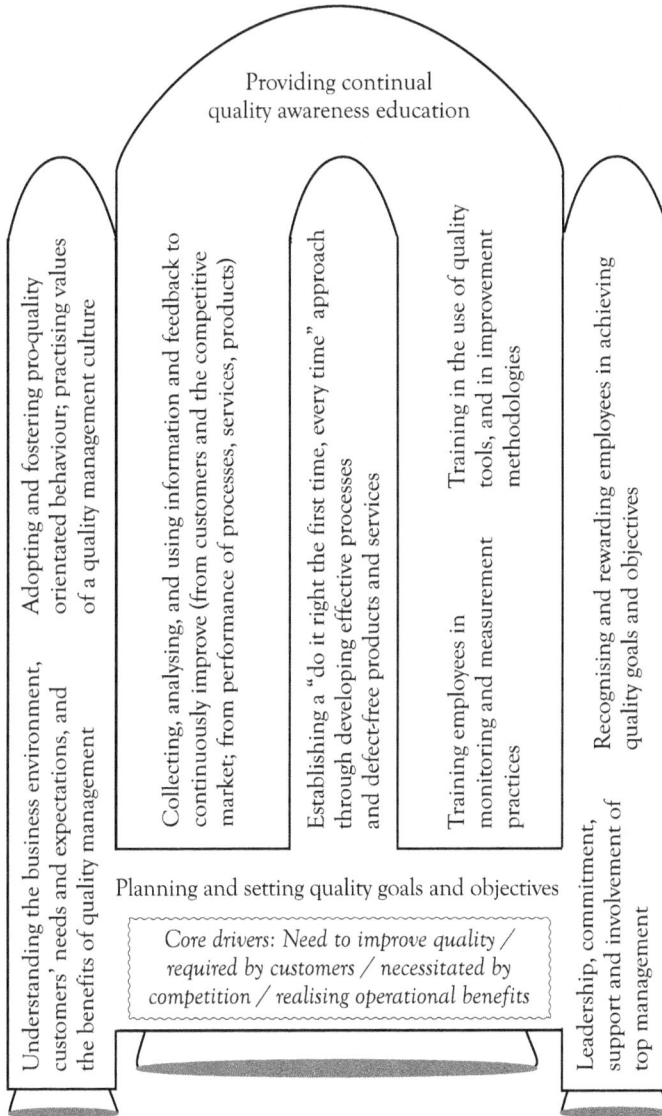

Providing continual quality awareness education

Adopting and fostering pro-quality orientated behaviour; practising values of a quality management culture

Collecting, analysing, and using information and feedback to continuously improve (from customers and the competitive market; from performance of processes, services, products)

Establishing a "do it right the first time, every time" approach through developing effective processes and defect-free products and services

Training in the use of quality tools, and in improvement methodologies

Training employees in monitoring and measurement practices

Recognising and rewarding employees in achieving quality goals and objectives

Understanding the business environment, customers' needs and expectations, and the benefits of quality management

Planning and setting quality goals and objectives

Core drivers: Need to improve quality / required by customers / necessitated by competition / realising operational benefits

Leadership, commitment, support and involvement of top management

Figure 1-1: Excellence Evolution – Equipped for continuous improvement

The planning and setting of quality goals and objectives is preceded by top management obtaining an understanding of customers' needs and expectations and of the business environment, and an understanding of quality management and its benefits and advantages.

An essential part of the goals and objectives is establishing a "do it right the first time, every time" approach through the development of effective processes and defect-free products and services. This relies on the actions of collection, analysis and use of information and feedback to understand processes, identify problems, and apply correction, improvement and prevention. These actions are facilitated by way of training in the use of quality tools and in improvement methodologies.

Crucial to this approach is providing continual quality awareness education to all employees, and the adoption and fostering of pro-quality orientated behaviour, practising values of a quality management culture, discussed in Part 1. A most necessary element is recognizing and rewarding employees in achieving quality goals and objectives.

Equipped as described in summary, and with the commitment, leadership, support and involvement of top management – and involvement requires participation in the various activities to manage and improve product quality and quality-affecting processes – the improvement journey can be sustained.

Questions for Chapter 1

1-1: There is much to learn from organisations that have persevered and developed successful quality management practices of very high standards that really bring results. (a) Identify ten characteristics and understandings that are shared by such organisations. (b) Explain how or why each characteristic and understanding identified will support quality management and business success.

1-2: What could be some of the possible factors or detracting influences in organisations in China that impedes the realisation of the true intent of quality management and continuous improvement?

PART 1: THE MAKINGS OF A QUALITY CULTURE

Subjects, practices, concepts and techniques contained in Part 1

Chapter 2: Organisational culture and quality values
- Organisational or corporate culture
- Values of a quality management culture
- Issues and actions that could oppose quality management culture values

---

Chapter 3: Effecting a quality orientated culture
- Influence of managers, leaders and "quality champions"
- Values that influence actions
- Influence of Visual Management
- Effect of mind-set: "growth" and "fixed"
- Influencing behaviour using a Human Performance System Model
- Making operator skills visible through a Skills Matrix Chart
- Effect of communication on behaviour
- Negative effect of the reward/penalty approach
- Positive effect of "Management by Walking Around"

2. ORGANISATIONAL CULTURE & QUALITY VALUES

Organisational culture is an important aspect of a company as it influences employees' behaviour, motivation and performance. The organisation's culture – what its employees collectively value, the organisation's commonly held beliefs and expectations, how employees communicate and make decisions – defines the organisation.

The values of a quality management culture are embedded in the organisation's culture. A key to improving an organisation's business success and competitiveness lies in the organisation developing a strong quality management culture.

This chapter describes and discusses:
1. Organisational culture
2. Values of a quality management culture and issues encountered

2.1 Organisational culture

Organisational or corporate culture is that distinctive combination of shared practices, values, beliefs, customs, work styles and relationships that develops within an organisation and guides the behaviour of its members. Organisational culture is shown in the following:
- The way that employees are treated and customers are handled.
- How freely information flows through the organisational structure.
- The extent to which freedom is allowed in decision-making.
- How employees are rewarded through identifying, owning, and solving problems.
- How initiatives, new ideas and ongoing improvements are encouraged.
- How motivated employees are towards trying to achieve the goals of the organisation instead of their own goals.
- The things that top management and leaders pay attention to and measure and control.
- Through the actions of employees living up to the brand image vision of the organisation.

Organisational culture affects the organisation's productivity and performance, attitude towards quality management, customer service, product design and its quality, and safety, attendance, punctuality, and concern for community and environment. It also influences whether the constant review for effectiveness and improvement is welcomed. Organisational culture is unique for every organisation.

In the West it is understood by the better organisations that organisational culture, and the management of this culture, is an important factor that has a major bearing on quality management and continuous improvement. Organisational culture is styled by the norms and values that an organisation chooses to follow, and top management plays a crucial role in shaping these norms and values.

China has a rich and unique cultural heritage which influences the culture within Chinese companies. Typically, the organisational cultural values are characterized by centralized decision-making, a paternalistic style of management, acceptance of hierarchy, and group orientation; they concern duty and obedience, labour discipline, and stress the importance of relationships (guān xì), "face" (miàn zi), and harmony. These cultural values greatly affect quality management.

Where top management and the leaders within an organisation place their attention is a powerful way in which values, beliefs and priorities are communicated. What leaders emphasize and measure become powerful ways of communicating messages, especially when the leaders are consistent; it is the consistency that sends the message about leaders' priorities, values and beliefs. If top management consistently stresses the importance of quality-related actions, and follows through to ensure that the actions are being carried out, and take an interest in the effectiveness of the result, then it is almost certain that quality-related actions will become a normal and a natural part of the company.

Managers appreciate that the business world is a dynamic environment, and that their organisation must be able to adapt to new demands and new initiatives. Case studies reflect the demise of companies that have not done this; these companies have simply not been able to survive in tough economic times. Modern quality management requires that a work style practice is taken of constant examination and review for effectiveness and for improvement; and it insists that corrective, preventive, and improvement action is taken at every level in the organisation. These quality management behavioural norms are promoted, supported and encouraged in world-class companies.

2.2 Values of a quality management culture and issues encountered

Many successful business transformations have occurred when the organisation has adopted the values of a quality management culture.

In the 1970s through to the 1980s, Jaguar Cars suffered from poor build quality; the sale of Jaguar vehicles suffered while German and Japanese vehicles gained market-share.

Jaguar Cars was worth £300,000,000 when Sir John Egan[1] took over as Chief Executive in the 1980s. He was very concerned about what his customers had to say about the build quality. Underdeveloped designs were being rushed into production, outdated processes were being used that had not benefitted from review and improvement, co-operation and teamwork within the workforce was poor, and bought-in component quality was questionable. He set about addressing all of these issues.

His steadfast commitment to his beliefs about what was important to the customer brought about a set of new values to Jaguar Cars, values that caused a quality transformation not only within his company but also within the companies supplying product to Jaguar Cars. The build quality gradually improved to establish Jaguar as a quality brand.

---

[1] Sir John Egan (born 1939), British industrialist associated with automotive, airports, construction and water industries. He was chief executive and chairman of Jaguar Cars from 1980 to 1990, chairman of Jaguar plc from 1985 to 1990, chief executive of BAA from 1990 to 1999, past president of the Confederation of British Industry and chairman of Severn Trent.

Less than a decade after Sir John Egan had taken over as Chief Executive, Jaguar Cars was sold for £1,600,000,000.

A quality management culture exists, and can be sustained, when the organisation adopts certain values. Ten such values found in world-class organisations are identified in table 2-1 and explained in this chapter.

Table 2-1: Principal values of a quality management culture

1. Focus on the customer
2. Know and support your internal customers
3. There must be teamwork and co-operation
4. Customer satisfaction must drive key performance indicators
5. Concentrate on finding the root cause of the problem
6. Quality management must be fact-based
7. Progress is made by way of solutions, not by finding personal fault
8. Everyone is involved
9. Quality management is integrated with overall management
10. People are the most important resource

Value 1: Focus on the customer

Keeping focus on the customer during the daily routine of work is an important value of a quality management culture. This value is put into action through treating the customer with courtesy and respect, communicating with good attitude, and responding promptly to enquiries and complaints. This must be applied from initial contact and order enquiry, through contract review, contract amendments, production surveillance conducted by the customer, product acceptance and handover, servicing and customer complaint handling.

Information to improve products, delivery, and services is obtained from feedback from customer experiences, analysis of customer complaints, and analysis of customer satisfaction survey results.

From personal experiences, the hospitality and courtesy shown by customer-facing personnel in Chinese industrial manufacturing concerns is exceptionally good. The order enquiry and contract agreement process is handled with proficiency. However, it seems that after this point, when the customer order enters the "realisation process" in the greater workings of the organisation, the "focus on the customer" value becomes overshadowed by prevailing Chinese culture values mentioned previously, notably, the importance of relationships within the organisation, duty and obedience, and harmony.

When the product does not arrive as per the delivery schedule in the contract, the customer is likely to label the Chinese organisation "over-promise, under deliver". To add to the customer's frustration, when the product eventually arrives, it arrives with quality problems. This is such a big concern that the customer resorts to sending his quality expert to the manufacturing organisation in China, sometimes to perform production surveillance, usually to monitor and witness product acceptance testing.

The Chinese cultural values of the importance of relationships within the organisation, duty and obedience, and maintaining harmony are of course most important, and the leader has a big responsibility to ensure that the "focus on the customer" value is maintained. However, take the scenario in a manufacturing organisation in China of an employee who is busy processing a most urgent and important customer issue, and the leader approaches the subordinate employee with a personal work request. Without hesitation, the employee will most likely change priorities and take up the leader's personal work request. No matter that the task may be of relatively minor urgency, because the request comes from the leader, that subordinate employee will in all probability immediately swing his attention to the requirement of the leader. It seemed to be inculcated into the subordinate employee that this was the proper order of things.

This thinking and consequent response places a huge responsibility on the leader's shoulders to ensure that the subordinate is rather directed in the business context of "focus on the customer".

Given a similar scenario in a well-managed Western organisation, the subordinate employee receiving the request or instruction will politely explain the situation to the leader, and prioritization of the leader's requested work will be determined through a quick discussion. The urgent and important customer issue will receive the higher priority.

Value 2: Know and support your internal customers

From order intake to delivery to the customer, there are numerous transactions between the people, functions and processes of an organisation. Internal customers are those employees provided or supplied information, material, services or product by other employees (internal suppliers). The internal customers are affected by how well the internal suppliers perform their job and what respect is given to urgency requirements. The extent to which each internal supplier supports their customers, i.e., meets their needs, determines the quality of the finished product and quality of service offered to the ultimate external customer.

The internal customer has specific requirements that must be met by its internal supplier in order to produce the output that meets the needs of its customer. Identification of these requirements is done in consultation and discussion, and the internal supplier actively strives to fulfill these requirements. This way of working prevents sub-optimal performance in the process-chain (discussed in Chapter 8.1).

Within organisations in China there appears to be room for improvement in the communication between departments; it was often found that there was a certain amount of "throwing the file over the wall". For instance, too often new design drawings were issued without taking into account the suitability of the production equipment or skills needed by the operator. Sometimes tolerances were specified excessively tight, to the extent that the tolerance could not be consistently maintained by the machine, i.e., the machine's process capability (discussed in Chapter 10.2) had a natural variance that exceeded the specified tolerance, on other occasions the key process points had not been explained to the operator. Issues such as these would cause unnecessary hardships and rejects during production.

Quality problems were frequently compounded by the lack of urgency of supply and support departments (internal suppliers) to respond to issues. Production was found doing "extra operations" to make good supplied product containing quality imperfections and defects. Also encountered was the situation where product, not conforming to drawing or process specification, would be quarantined for several days waiting the design engineer's evaluation; during the several days, work-in-progress would build up and knock-on effects down the line would cause the product to be shipped late to the customer.

Value 3: There must be teamwork and co-operation

Teamwork and co-operation is a most necessary value of quality management. Teamwork is used to achieve results and to bring solutions; it is known to create synergy, i.e., the output of the team is greater than the sum of the individual inputs. Teamwork is an integral part of continuous improvement.

Teamwork requires a certain mind-set where there is no power-play, where each team member is regarded as a contributor to achieving a successful result. Co-operation is required at an individual and inter-department level; this entails accessing needed resources and appropriate information.

Functional boundaries that inhibit co-operation and the flow of information are actively eliminated in Western organisations, and needed information is readily accessible. In large manufacturing organisations in China, functional empires are strongly entrenched which tends to impede cross-functional teamwork and co-operation; information is often not readily available and even exclusive to a certain person. For example, when seeking work-related information needed to sort out a serious and urgent quality problem, it was often found that this information was not available because a certain person was not available, and that the data was locked in this person's desk or office, and that the individual or team needing the data had to wait, sometimes for several days, for this person to return. Clearly, this action has not supported co-operation to solve the problem and has impeded the team's progress.

Behaviour traits of preservation of saving face and self-preservation thinking are prevalent to varying degrees in employees all over the world and can be affected by culture and the working environment.

In a well-managed organisation, if the skilled employee's personal work process produces defects and he is unable to sort out the problem, he is encouraged to call for assistance; through co-operating with colleagues, inspectors and engineers, the desired result or outcome of his work and therefore of his work unit can be achieved.

If the organisation has a blame-orientated culture, the employee would very likely be afraid of accepting personal accountability for a poor result or bad outcome. Behaviour would then be predisposed to protection of self rather than towards the benefit of the individual's unit or the organisation.

Behaviour traits of preservation of saving face and self-preservation thinking are accentuated by Chinese culture and therefore particularly acute in employees in organisations in China. In the case of the skilled employee's personal work process producing defects and the organisation is operating a reward/penalty scheme, the employee will quite likely cover up the issue rather than expose it by asking for help – and this type of reaction will be more pronounced in an organisation having a blame-orientated culture.

Value 4: Customer satisfaction must drive key performance metrics

The customer is the lifeblood of the business. Unhappy customers have no qualms about taking their business elsewhere, and with their defection come loss of company earnings.

A customer who uses your product and experiences your service has certain needs and expectations in mind. If the product and service meets or exceeds those expectations, then, in the mind of the customer, you have given him a quality product and a quality service. Quality therefore relates to the customer's perception. Customers compare actual performance of the product and the service experience to their own set of expectations, and reach a judgment.

The judgment of quality is not confined to product conformity; the customer's judgment includes the entire experience from initial contact with the sales-person, through contract agreement and technical discussion, to delivery, installation and after sales support. This is well understood in successful organisations. Information is routinely sought relating to customer satisfaction to identify weaknesses, opportunities, threats and strengths. This information is analysed to identify trends and issues requiring action.

A best practice is for Key Performance Indicators to be established of customer satisfaction from metrics such as,
- response time to answer customer queries,
- performance against agreed delivery schedule,
- reduction in the number of customer complaints,
- time to handle customer issues to satisfactory resolution,
- evidence that show the company is listening to its customers (e.g., through customer surveys),
- growth of number of customers,
- number of customers that are repeat buyers,
- reduction of the cost of poor quality,
- elimination of a specific reliability affecting design weakness.

Customer satisfaction indicators such as those above are continually monitored, and timely action is taken to make improvement.

In industrial manufacturing organisations in China, findings in general indicated that the scope of key measures of customer satisfaction could have been improved. Those that were obtained could have had their importance emphasized by way of performance goals, or had some value extracted through appropriate analysis of collected data.

For example, importance could have been emphasized in the response to handling customer complaints by defining a time duration goal for the time taken to achieve satisfactory resolution, and emphasis of the gap between actual and achieved performance against agreed delivery schedule could have been stressed instead of emphasizing production quota, and a target response time could have been placed on time to answer customer queries to emphasize this importance.

Value 5: Concentrate on finding the root cause of the problem

The integral part of quality management, which is quality improvement, requires a questioning attitude – using the 5 Whys question-asking technique (see Appendix 1) to determine the root cause of a defect is very effective – and this questioning attitude is best coupled with a long-term approach. With a long-term approach emerges an obsession with the prevention of reoccurrence of the problem or shortcoming, and the seeking of incremental gains. This type of thinking is at the heart of KAIZEN and leads, over a period of time, to actions which result in improved process control, improved product conformity, improved product designs, improved service operations, etc.

The time that it takes for this type of improvement thinking to develop will depend upon factors such as leadership, employee involvement, persistency or constancy of purpose, and management complexities.

Improvement thinking applies when thinking about what corrective action should be taken, and it applies when defects are routinely analysed for trends, customer complaints are analysed to identify what action must be taken to avoid a repeat, and when internal quality audit results are reviewed to determine how processes can be more effectively applied or how they can be improved.

Management wants quick results and in organisations in China it was not uncommon to find a short-term quick fix "fire-fighting" mentality. Action as a result of continual analysis was too rarely in evidence. Attempts by top management to get the PDCA method and way of thinking established were seen (Appendix 1 and Chapter 8.2 contain explanations of PDCA) but many middle-managers appeared reluctant to take the initiative – perhaps they did not want to disturb the status quo or risk disturbing harmony. Top management's intervention appeared to be a necessary requirement to the solving of significant problems but getting to the root cause was not easy and it was noticed that often, personnel tasked with solving problems tended to "recognise" causes that seemed to support popular or powerful beliefs. This somewhat prejudiced actions which did not usually lead directly to uncovering the causes that needed to be addressed.

Value 6: Quality management must be fact-based

In quality management "facts and data" carry huge value. Quality can only be managed with facts. Ongoing improvements can only be made to products and processes by collecting and analysing facts and data. Data is needed to establish quality status and situation and is analysed to produce information which is used to direct appropriate action.

If management does not have the right measures applied to a situation, it may never be known whether the situation is under control. The design engineer needs his design validated in terms of performance, safety and reliability and judgements and improvements can only be made based on facts and data gleaned from the likes of testing and proving programmes. In production quality control, facts and data are needed to determine process variability and process settings, and to take action to reduce defectives and to eliminate repeat problems.

Data collection in industrial manufacturing organisations in China appeared to be biased towards things deemed important by top management, and a lot of importance appeared to be attached to labour control and discipline. For instance, data was collected on the number of final inspection rejects and scrap produced and used to calculate penalties; the thinking was that giving penalties to people would work to control bad quality – but this approach assumed that rejects and scrap were caused by operator error or operator negligence.

In fact, the cause of poor quality could lie elsewhere, as illustrated in figure 2-1 where some of the many possible causes of poor quality are grouped under categories of work performer, machine, material, management, method and environment.

Cause: Management
- Failure to identify or understand customer needs
- Poor communication of quality requirements
- Lack of understanding the process approach
- Inadequate or poor planning
- Lack of supervision and process monitoring
- No adequate emphasis on quality measurement
- Lack of fact-based decision making
- Lack of recognition system and flawed incentive scheme
- No measurable quality included in Key Point Indicators
- Bias and/or favouritism
- Closed attitude towards change
- Lack of decision making

Cause: Material
- Excessive variation
- Impurities or contaminants
- Incorrect material grade
- Unclearly specified material
- Unsuitable material type

Cause: Method
- No instructions for critical-to-quality processes
- No clear work instructions
- No clear process procedures
- Too rigid or too relaxed requirements or specifications
- Conflicting requirements

Effect: Poor Quality

Cause: Environment
- Unsuitable lighting for type of work
- Unsuitable temperature
- Poorly controlled temperature
- Improper humidity
- No protection from static electricity

Cause: Machine
- Limited machine capability
- Improper set-up and/or calibration
- Lack of maintenance
- No replacement parts
- Excessive wear
- Unsuitable or outdated technology

Cause: Work Performer
- Lack of skills
- Lack of training in quality and work practices
- Not following procedures or work instructions
- Unsuitable personnel
- Personnel taking shortcuts
- Lack of motivation or interest
- Fear of failure
- Excessive stress
- Shortage of work performers

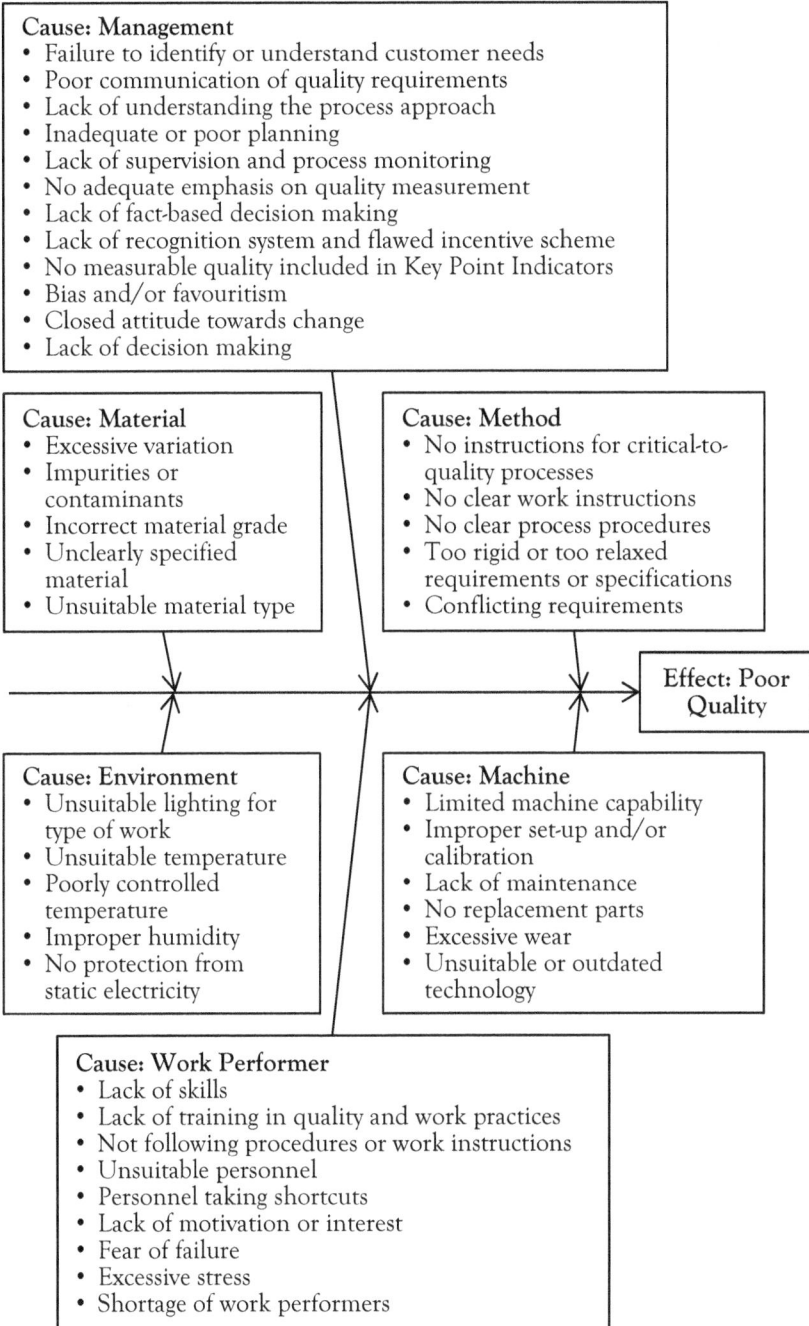

Figure 2-1: Possible causes of poor quality

A lot of wasted effort and misdirected action was witnessed due to the tendency to default to guesswork. Perhaps this was due to employees "recognizing" causes that were politically correct or that supported powerful beliefs, as previously mentioned, or because there were no specially trained quality personnel. The action taken was invariably based on some historical assumption, and more often than not the operator was factored in as the assumed weak link.

Trained quality personnel would be able to show what data should be collected; they could design quality related data collection methods and arrange for the data to be captured and analysed.

It is felt that much long-term benefit can be derived from pursuing fact-based problem solving and rejecting quick fixes not supported by data.

Value 7: Progress is made by way of solutions, not by finding personal fault

The attitudes toward problems and actions taken in handling problems are important concerns in a quality management culture. Finding solutions to problems, without trying to place the blame on someone, is intrinsic to a quality management culture.

Continuous improvement is given impetus and energy when the focus moves to the calling of attention to solve the problem and if the problem is caused by the individual's actions, then attention needs to be directed to actions to modify the individual's behaviour. Continuous improvement is stifled when problems are personalized and individuals are repeatedly threatened or shamed; this promotes a blame culture.

Certainly in the better organisations in the West it is long known that employees will not willingly speak freely when a blame culture is prevalent, let alone admit to a mistake. Employees in this blame culture will avoid collecting data, and avoid putting their name to documents that could result in the work being traced back to them. A blame culture is avoided as this is known to be non-productive, to promote the hiding of issues and the non-disclosure of certain facts, and even to cause employees to be dishonest.

A general observation was that in industrial manufacturing organisations in China there appeared to be some evidence of a blame culture. Management appeared to be quick to assume that the fault was caused by careless workers and believed that punishment would cause the workers to work more carefully. This approach introduced an element of fear and most likely caused the workers to be reluctant to disclose problems which included quality-related problems. This tended to make the obtaining of accurate facts a headache.

The most necessary requirement to finding solutions is for employees to be allowed to have a questioning attitude, and be allowed to attempt to find solutions. But this requires management to allow and encourage this, and to be non-judgmental, treating the employee justly and fairly.

In the better Western organisations it is stressed, as a sensible business point of view, to focus on finding solutions. Employees are encouraged to be open and honest about quality-related problems. Every attempt is made to remove blame or shame associated with a quality-related problem. If it is found that the cause of the problem relates to worker ability, the solution is to address the ability of the worker through suitable means such as on-the-job training.

Value 8: Everyone is involved

In their everyday work, everyone is expected to contribute in some way to "making quality happen". This contribution is by way of the actions of personnel such as the following in specifying or clarifying or reviewing, checking, measuring, analysing, feeding back or improving:
  - managers, supervisors and clerical staff,
  - sales and marketing personnel,
  - design, process and quality engineers,
  - purchasing and sub-contract management personnel,
  - production planners, supervisors, operators and inspectors,
  - test, commissioning and product acceptance personnel,

- stores, dispatch and delivery personnel,
- field service personnel and field agents,
- customer service personnel,
- accountants and human resource personnel

In the best organisations in the West, wherever possible through work design, "making quality happen", and the associated rules are built into each job; examples are as follows:

- Managers are given quality improvement objectives as part of their "Key Performance Indicators" , such as:
  - improve the contract review process with regard to checking that the company can achieve customer stated requirements
  - increase effectiveness of the production readiness design review
  - improve "Operator Self-Control"
  - reduce time-to-deliver
  - reduce repeat problems
  - improve skills competency of workers
  - improve quality awareness of employees
  - increase the benefit of internal quality audits
  - improve the value of customer surveys
- Sales-people are required to ensure that the customer's requirements are reviewed prior to acceptance of the order. A contract review process is employed to check whether the company has the capabilities to meet requirements of product design, manufacture, delivery schedule and service.
- The design team is required to check the "quality of design" during the stages of product development and a cross-functional design review process is followed with objectives to:
  - control risk, i.e., risk of an inadequate design with respect to meeting customer's requirements and excessive cost;
  - ensure satisfactory performance, safety, reliability, maintainability and interchange-ability with existing assemblies;
  - ensure ease of produce-ability, installation and servicing.

It is usual to find other departments such as sales, process engineering, production, and quality playing an important part in ensuring the "quality of design".

- Process and quality engineers are required to prepare the means to ensure "quality of conformance". This includes,
  - specifying suitable production and measurement equipment;
  - specifying production, inspection and test methods;
  - identifying critical-to-quality parameters that require attention;
  - preparing process and quality control instructions.
- Production management and supervision are required to check that the means needed to achieve "quality of conformance" and "Operator Self-Control" is provided, and, in conjunction with relevant parties, ensure that job training is given where required.
- Operators are required to practice "Operator Self-Control" to ensure "quality of conformance" by making certain that what they are given is adequate, and by checking that their work is in conformance.
- Order preparation and stores despatch personnel are required to check against the agreed customer order that despatch address as well as product particulars and quantity is correct, and that packaging is appropriate to prevent deterioration of quality.
- Commissioning and field service engineers are required to ensure that "quality of conformance" is achieved in the delivered product, and, to enable improvement of product and service, they are required to give relevant feedback to sales, design, and production.

In Chinese industrial manufacturing organisations it was found that sometimes the work design did not allow for employees' actions to follow the activities necessary to "making quality happen", or employees actions did not follow the expected course to achieve this end. To cite some observations using the examples given above:

- Contract review was often done after acceptance of the customer order – this sequence detracts from its purpose.
- "Peer review" (where engineering drawings were checked by a colleague from the same department) was taken to suffice as a design review.

- Although it was claimed that operators performed "Operator Self-Control", there was generally an absence at their workstations of the necessary means to do so and an absence of routine in-process checking actions performed by operators.
- Customer orders were dispatched that were often incomplete and sometimes incorrect.
- Commissioning and field service engineers rarely fed anything of substance back that could be used to overcome difficulties in the field or to enable improvement.

This provides a glimpse of some of the issues where quality improvement actions and teamwork would shine through. Change to procedure or process would most often be required and this would invariably involve modifying the everyday actions of and transactions between people, functions or departments. But it was very difficult to find willing participants, particularly in the middle management group, to alter their ways to make the change, particularly to spontaneously act on cross-boundary feedback to make improvement. It seemed that many employees chose rather to "let things remain as they are". Sometimes it seemed that their actions demonstrated the notion "if we procrastinate long enough the person trying to make the change will eventually give up", and it seemed that this approach of theirs was unfortunately often successful.

Value 9: Quality management is integrated with overall management

Quality management has endured and evolved over decades. In most organisations in the West, quality management is now firmly integrated with overall management, and it is accepted as a fundamental factor in business success; there is no separate programme for quality management, and there is no single special department for carrying out ISO 9001 (GB/T 19001) quality management processes.

The Western organisations that have reaped benefits from quality management regard it as part of their overall management process.

Organisations that have mistaken quality management as a management programme or project have not reaped benefits.

What is the difference between a management process and a programme? Philip Crosby [2] explains the difference by examining successful and unsuccessful dieters. To a serious dieter, a weight-loss strategy is a process. When it is such, the overweight person changes their eating and exercise habits forever. He reaches his goals and maintains his desired weight. A dieter, who selects a diet programme rather than a process, may lose weight during the diet programme, but regains it later when he decides to end the programme.

A management programme is typically something with a beginning and an end whereas a management process is a methodology that is developed to replace the old ways and to guide corporate activity year after year.

It is unfortunate that some managers regard quality management as a programme or project. In some Chinese organisations that had attained GB/T 19001 (ISO 9001) accreditation some years ago, it appeared that a project goal to attain ISO accreditation had been set, and when it was deemed that this goal had been achieved, e.g., on being granted a GB/T 19001 (ISO 9001) accreditation certificate, development of the quality management system slowed to a practical standstill.

Quality management and ISO 9001 have continued to develop with huge changes occurring over the past 25 years. A major shift occurred in the year 2000 when the orientation of ISO 9001 changed from an inspection based standard stipulating conformance to requirements, to a management system standard emphasizing quality performance, continuous improvement, and prevention. Further improvements included the process-based approach. These changes have facilitated the integration of quality management processes with an organisation's overall management processes.

In Chinese industrial manufacturing organisations sporting GB/T 19001 accreditation, examination of their approaches more often than

---

[2] Philip B Crosby (June 18, 1926 – August 18, 2001), businessman, quality management consultant, and author

not revealed much emphasis on inspection and less emphasis on improvement as part of their day-to-day routines. There was also a general lack of quality problem prevention-centred thinking and unfamiliarity with the process-based approach. This suggests that further work is required to update the orientation of some GB/T 19001 implementations.

Management are very concerned about the overall effectiveness and efficiency of their management system practices and processes. For this to be achieved, the various processes, including those of quality management, need to dovetail and work together to produce the outcomes desired by the company. In industrial manufacturing organisations in China, however, it was generally not clearly apparent that the processes of quality management were accepted as an integral part in helping the overall management system produce the desired outcomes. This was firstly usually apparent in the annual management review (an ISO 9001 accreditation requirement) which appeared to be a summary of the once-a-year internal audit of quality management departmental procedures, and a focus on quality objectives pertaining only to non-conforming products in production.

A more balanced set of quality objectives would cover a wider range of areas of importance that work to achieve the desired outcomes of a quality managed company and would include objectives such as those relating to customer satisfaction and the risks of the product or service not satisfying stated or implied needs, continuous improvement, skills and competencies, supplier performance, and product conformance.

This possible poor understanding of quality management was also apparent in those companies that had embarked on what they referred to as operational excellence programs such as Lean Manufacturing; these programs often had their own Lean Office developing processes in isolation to those of quality management. In the West it is usual to find that quality management establishes a suitable environment for World Class Manufacturing initiatives such as Lean Manufacturing. It can be said that there is a synergistic partnership – quality management concerns effectiveness which is the extent to which the organisation's practices and the outputs of its processes meet the needs and expectations of its customers, and the intent of "Lean" is to identify and eliminate every form of waste[3] – the synergy lies in the fact that as waste is eliminated, effectiveness improves, and efficiency improves as evidenced by the reduction of through-processing time and cost, and so this partnership works to deliver to the customer quality conforming products on time at the right price.

To aid the overall integration of management systems, an increasing number of organisations in the West are taking advantage of the common structure, terms and definitions developed in the later revisions of ISO 9001 quality management systems standard, and ISO 14001 environmental management standard, and the OHSAS 18001 occupational health and safety management system standard, to ease incorporation of these into their organisations, and such that the common management system auditing standard, ISO 19011, can be readily used to simplify the auditing process.

Value 10: People are the most important resource

The organisations that are most successful at quality management treat their employees as their most important resource. These organisations have learnt that employees that have high quality consciousness and good knowledge of quality processes and practices, and that are trained, empowered and enabled to use this quality consciousness and knowledge, bring success to their organisation. Every employee takes responsibility for their own actions, and they check their own activities. Every employee is empowered to participate fully in the improvement process; this stimulates commitment and interest which leads to fulfilment and job satisfaction.

---

[3] Note that whereas effectiveness is having the right output at the right place, at the right time, at the right price, the broad understanding of "waste" in Lean Manufacturing is *not having* the right output at the right place, at the right time, at the right price.

Differences were encountered with respect to employee development, training, and empowerment; the main differences are as follows:

- Quality education programmes in the successful Western industrial manufacturing organisation is a continually running initiative. The number of employees receiving this education is monitored and reported. In most organisations in China, quality education programmes were ordinarily fairly sporadic.

- The skills and competencies of each key operator receive a lot of attention in Western organisations; a best practice is that of the performance of each key operator being regularly reviewed by his manager or supervisor to identify gaps in skills and competencies, and the taking of improvement action. For higher level knowledge worker personnel, a best practice is that of Key Performance Indicators (KPIs) being set to include quality, and for KPIs to be reviewed periodically to identify the means to improve performance. In organisations in China it was rare to find practices involving the review of operator skills and competencies, and KPI practice appeared to be new and not yet perfected.

- "Operator Self-Control" is standard practice in Western organisations. The better organisations make sure that;
  - means are provided for operators to perform quality self-control;
  - training is given so that operators can perform in a problem-prevention manner;
  - measures are in place for operators to detect and record problems.
  The expectation for operators to control quality was prevalent in industrial manufacturing companies in China but, in comparison, the level of support to enable "Operator Self-Control" was minimal.

- In successful Western organisations, top management takes the leadership on quality management and middle management follow though in full support. In Chinese organisations, the nod of approval from top management commanding its use proved to be essential for middle management to then follow the procedural quality routines. Substantive quality management actions only tended to happen when top management personally took interest and followed through the actions that they initiated.

In many industrial manufacturing organisations in China it is not unusual to find significant capital investment in buildings, advanced manufacturing equipment and machinery. There is an expectation for this equipment and machinery to consistently produce quality products. However, there is often no evidence that the means necessary to support the operators to consistently make quality product have been taken in account, or, if they have been provided, there is no evidence of them being routinely practised.

For instance, the operator with minimal engineering education may be given a complicated engineering drawing with no indication of what to check or no indication of the criticality of a dimension or characteristic, he may not have at his workstation necessary instruction on how to control the quality with standard operating procedures or job inspection and measurement routines, he may not be given made-for-purpose measurement apparatus, and quality control practices such as quality status identification may not be in evidence; i.e., the effective means to support the operators for "Operator Self-Control" were rarely to be seen where needed.

The organisations that are most successful at quality management make a habit of acknowledging their most important resource – their people; they especially give recognition to those individuals and teams who distinguish themselves in achievements and contributions that directly contribute to the key priorities and values set by the organisation. These companies know how important it is for their employees to be recognised for their excellent work by their management, peers, friends and family; recognition nourishes the spirit and promotes the quality process. Recognition can be effectively conveyed, for example, by way of top management presenting trophies and certificates.

Company-internal recognition processes for contribution to quality, as seen in the best quality managed Western manufacturing organisations include, for instance:

- For product quality – awarded to each unit, workshop, and even to each cell (as in Cellular Manufacturing) – a star rating following a monthly standardised product quality assessment. Probably the most effective display of star rating is as seen on a graph that indicates, over time, the improvement in product quality or maintenance of a high consistency in product quality or otherwise.

- For overall quality – awarded to an individual or a group of people on a yearly basis – a Customer Focus Quality award. This significant award is most often accompanied by a symbolic trophy or plaque whereupon the names of each year's recipients are engraved. It is usually placed on display in a prominent place in the organisation such as the company's reception area.

Recognition of employees was seen to take place in China (e.g., best 6S unit) but no company-internal recognition processes were in evidence that acknowledged achievements in quality per se.

Questions for Chapter 2

2-1: Organisational culture is that distinctive combination of shared practices, values, beliefs, customs, work styles and relationships that develops within an organisation and guides the behaviour of its members. Present a brief summary of practices, values, beliefs, customs, work styles and relationships that affect the behaviour, motivation and performance of an organisation's members.

2-2: Chapter 2 describes ten values of a quality management culture. These values enable and support aspirations of a business organisation such as the securing and maintaining of customer satisfaction, the obtaining of a high level of efficiency and effectiveness of internal operations, and the achieving of continuous improvement to processes, products and services. Write a short essay on the manner in which the ten described values enable and support the aforementioned aspirations.

2-3: Provide examples of possible actions that could oppose each of the ten principle values of the quality management culture discussed in Chapter 2.

3. EFFECTING A QUALITY ORIENTATED CULTURE

Bringing about a highly favourable quality management culture takes a lot of time and perseverance especially in a large organisation with its different organisational levels. Changes in emphasis and attitude are needed at all levels which involve a different way of doing things.

There are many factors, influences and actions in an organisation that have a bearing on bringing about a quality management culture; prominent among these, the following are discussed in this chapter:

1. Influence of managers and leaders
2. Living the values through actions
3. Visual Management as a culture influencing agent
4. Attitude and mind-set affects action people take
5. Influencing and encouraging behaviour
6. Effects of communication

3.1 Influence of managers and leaders

The organisation's leaders and managers have a huge influence on a company's quality management culture. Their attitude, drive and passion, and how they engage with the organisation's employees sets the stage.

Top management has been seen trying to improve the quality culture by describing to their employees the new emphasis and attitude they seek. While some dutifully conscientious employees will act in ways they think fits the description, most of the employees carry on as usual – perhaps this is because it sounds to them like another management pronouncement. However, when top management stipulate the specific actions that are required, and introduce frequent measurement and monitoring, this directly focuses the employees and changes the actions that they take. Moreover, by repeatedly doing the actions, the employees' behaviour is influenced. Over time the new behaviour causes a change in attitude and emphasis.

Responses to the actions need to be regularly monitored by the organisation's managers and leaders, and follow-up actions taken as well as adjustments made as necessary; persistence is required because people will probably repeatedly relapse back into their old habits or behaviours before the new behaviour becomes engrained as the new way of doing things.

"Quality champions", play an important part in bringing about a quality management culture. These are most likely to be senior-level personnel in prominent positions in the organisation that comprehend quality management, and have quality education at least as outlined in table 7-4, and that have a progressive attitude. (See "growth mind-set" in Chapter 3.4.) They are highly motivated and have the presence and ability to engage with employees at all levels, and the know-how to remove organisational and personal barriers to progress.

Senior-level "quality champions" maintain momentum of action; their thinking, approach, knowledge of quality tools and techniques, and their enthusiasm and perseverance to overcome objections and obstacles is important for ultimate success.

3.2 Living the values through actions

The following narrative illustrates that the values of a quality management culture are effectively used to make progress in quality improvement, and are instilled through problem solving and continuous improvement actions.

Due to the increase in demand for mid-size lifting chain, management decided to re-commission an old set of "semi-retired" machines to produce this chain.

Lifting chain is safety critical and the mechanical properties are very important, among which is the consistency of tensile strength of the chain-links. From the outset, the lifting chain produced on this set of machines suffered from a high occurrence of tensile strength chain-link failures. The scrap rate was extremely high. The links appeared to break at random points in the chain length. Now and then a batch containing some marginal links would slip past the factory screening tests and end up being delivered to the customer. The number of customer complaints grew.

Top management were unhappy because the production failure rate was extremely high, and above all, the risk to the customer and the business was worrying. In their discussions they expressed doubt over the work of the operators. The production line-manager told his operators that they had to work harder to improve the consistency of the quality.

As the weeks went by, despite constant nagging to improve, nothing changed. The operators seemed to pay no attention to management's concerns. Management labelled the operators as having a bad attitude. The tension between the operators and management grew. The chain-link failures remained high.

The newly recruited quality engineer began an analysis of the problem. He observed that the problem was prevalent in both day-shift and night-shift production runs with slightly fewer tensile test failures in evidence during the day-shift. "Operator Self-Control" was practiced and he noted that the night-shift operators made slightly less use of the go/no-go gauges that they were provided than did day-shift operators.

The quality engineer met with both day-shift and night-shift operators to discuss the problem of the high number of chain-link failures. These operators were experienced chain-makers, but the quality engineer soon realised that nobody had explained to them how important their work was – they were controlling machines that directly affected a critical-to-quality characteristic of the product. He explained to them how their work directly impacted on failure to meet a critical user requirement and led to unsafe in-service conditions.

He found the operators very disgruntled with management. They were angry and thought that management were unreasonable to expect high volume and consistent quality chain from the old machines. They believed that management did not respect their efforts. The day-shift operators believed that the higher failure rate occurring during the night-shift was because the night-shift operators were not as good at the job as they were. Fingers of blame pointed everywhere but not to the most probable causes of the problem.

The quality engineer agreed that old and worn machinery could be a cause, and operator skill had to be factored into the situation, however he explained that there were many variables that could be causing the problem (see figure 2-1). To find the cause would need the collection and analysis of data.

He stressed that their work was highly valued and asked them to help to establish the cause of the problem. He asked them to take measurements of the length of the formed link before welding, and the length after welding. He asked them to measure five successive links in a continuous production run every 15 minutes. Some explanation soon had the operators understanding that the purpose of the exercise was to establish the nature of process performance and its variability. (In Chapter 10.1 and 10.2, process variability is discussed, and in Chapter 11.7, control charts are explained.)

Analysis of the graphically captured process performance on control charts showed a distinct trend – the length of the formed link after welding steadily increased while the length of the formed link before welding remained unchanged. This meant that, in the upset butt-welding operation, the amount of weld-upset decreased steadily over time. Insufficient weld-upset was causing the chain-link failures. The insufficient weld-upset limit was neared only after 30 minutes! The operators explained that they had observed how quickly the copper welding electrodes deformed under pressure and wore down.

The quality engineer asked the metallurgist to ascertain the variation in the hardness of the chain steel (which was die-drawn to a specified size on site), and to research and recommend copper welding electrodes that were harder and more resistant to wear. The variation in the hardness of the chain steel proved to be slightly excessive and steps were put in place to reduce this variation (arising during the die-drawing process). Harder and more wear resistant copper welding electrodes were sourced.

Once these changes had been put in place, the in-process measurement exercise was repeated. The analysis showed that the length of the formed link after welding still steadily increased but the insufficient weld-upset limit was now reached after 90 minutes of chain production – a great improvement over 30 minutes.

The operators learnt how to interpret the process control charts, and that they had to adjust the welding machine every 80 minutes to avoid insufficient weld-upset. They learnt the need to use the go/no-go gauges more frequently. Soon this new routine of measurement and adjustment was established. The operators were happy with their new found control chart skill and discussed this among their colleagues who expressed an interest in its further application to their processes. The internal supplier (die-drawing) became more aware of his internal customer's (chain forming and welding) requirements and took measures to maintain consistent hardness. The metallurgist earned respect from his contribution.

The chain consistently passed tensile-strength tests. Rework and scrap costs reduced, and customer complaints dropped. Management were delighted with the results. The small increase in overall process time to accommodate the more frequent welding machine adjustments was fully understood and accommodated.

The approach, incorporating the values of a quality management culture previously described in Chapter 2.2, was applied throughout the factory. The quality engineer arranged for quality information to be placed on display boards located in the factory. The quality information concerned graphical displays and charts. These showed type and number of defects, and indicated prevalent problems, corrective actions taken or planned to be taken, and results. The displays were updated weekly and the information was discussed thus engaging internal suppliers, internal customers, and top management. Momentum was sustained.

Typical issues that can be overcome through living the values through actions:

Progress in developing a quality environment will be made through encouraging problem solving and continuous improvement actions, and, in the "doing" part by drawing in and mobilizing the values of the quality management culture described in Chapter 2.2.

Some issues encountered in industrial manufacturing organisations that can benefit from this approach are:

- The attitude of "It's not my problem". This is when workers, engineers and middle-managers point the "finger of fault" to another, and avoid getting involved. (See values 7 and 8.)
- Difficulties in collecting quality information from live operations. In the workplace, there can be a general resistance to recording and reporting defect information. (See values 3, 5, 6 and 7.)
- Actions not being followed through unless personally "walked through". This can be particularly prevalent in corrective actions arising from internal quality audits where delays of many weeks have been noted to occur in waiting for response, despite continual reminding. (See values 3 and 9.)
- People "giving up" asking. This is found when repeat problems, typically such as those encountered during assembly, are repeatedly reported for many months but no corrective action is attempted, and the persons reporting the issues are told to stop being a nuisance. (See values 3, 7 and 8.)
- Customer complaints being unresolved for many months because of Production not allocating a replacement part from their set production quota. (See values 1, 4 and 8.)
- The internal customer accepting components from an internal supplier that do not meet his requirements. A number of longstanding issues that have been identified arise, for instance, from an inadequate specification or sub-standard finish causing the internal customer to set up a special operation to improve the quality of received items to an acceptable level. (See values 1 and 2.)

3.3 Visual Management as a culture influencing agent

Visual Management utilizes the human senses to communicate quality information, which helps people understand situations at a glance.

Visual Management is a culture influencing agent; it facilitates order, control of quality and problem prevention.

Quality information charts placed on display boards are part of Visual Management. In the example given in Chapter 3.2 the displays proved to be very effective because the information was relevant and meaningful, continually updated, easy-to-understand, discussed and acted upon. Other types of Visual Management include travelling job cards, process control cards giving work instructions and indicating status of product, discrepant material notices, quarantine labels, and also floor markings, tool boards, and various signage to remind people to wear personal protective equipment, where to place non-conforming product, or where to find process information.

Visual Management comprises of visual displays and visual controls.

Visual displays are such as charts informing of product yield and conveying information to people in an area, graphs showing trends of certain quality issues, and pictures illustrating quality defects.

Visual controls are such as signs, labels and markings, and are intended to control or guide the actions of people in an area. They are primarily employed for safety, process and quality control purposes; floor-markings and signs may be used to control through-ways and where to place components, e.g., before and after different types of processing or that are awaiting test and inspection, and, a clearly labelled tool storage board will indicate where a tool belongs and what tools are missing from the board.

The Chinese people employ the visual elements very well. Manufacturing organisations in China were noted to use eye-catching banners and displays, however, not often seen in use were visual displays and visual controls concerning quality control or improvement.

Various approaches to encourage the use of Visual Management in quality management have been personally tried. Regrettably, generally, the use has been maintained for a limited period, or use seems to have been regarded as a one-off "special occasion" event. Production process control cards have not been maintained, discrepant material notices have lost their meaning and analysis-value due to lack of declaration of information, and labels on material-in-process that indicate the status of material (e.g., in quarantine), have been abandoned. Graphical displays of quality performance that should have been updated regularly, to maintain their relevance, have remained stagnant.

When problems, abnormalities, and deviations from standards, or absence of controls are made visible, action can be immediately taken. It behoves the organisation's management to make Visual Management work.

Visual Management has the power to be a quality management culture influencing agent; it is described as serving functions of transparency, discipline, continuous improvement, job facilitation, on-the-job training, creating shared ownership, and management by facts, simplification and unification[4]. Each of these functions is explained in table 3-1, along with examples of alternative practices that will most probably be used in the absence of Visual Management; the alternative practices are likely to promote behaviour responses that are inclined to give rise to discord and an undesirable quality management culture.

Table 3-1: Visual Management functions and alternative practices

Explanation of Visual Management Functions	Alternative Practices
Transparency: Activities, components, tools and indicators of output and quality performance are in sight. The status of the workplace can be understood at a glance by the manager, engineer and worker.	Work information kept in personal notes and personal computers. Measurement tools locked away or in a place remote from the workplace.
Discipline: Controls that range from subtle pressure to influence behaviour by way of visual elements, to mistake-proofing devices.	Threats, scolding, giving of punishment and demoting.
Continuous Improvement: Visual display tools – that display the problem and communicate	Static organisations relying on big improvement leaps

---

[4] Visual Management, Algan Tezel, Prof Koskela, Dr Tzortzopoulos. University of Salford, 2009 and 2011.

suggestions such as an idea board; that communicate problem solving techniques, actions and results such as Basic Quality Tools and that praise involvement effort such as a star board.	through considerable investment. (See Appendix 1 for an explanation of static and learning organisations.)
Job Facilitation: Visual aids or visual quality standards used to help people to perform their jobs. Also, when the amount of information required to complete a task pushes the capacity of the working memory, it can be made available through visual displays.	Expecting people to perform well at their jobs, especially jobs requiring a lot of information, without providing them any aids.
On-the-Job Training: Visual elements are useful in imparting "know-how" or tacit knowledge that is difficult to transfer to another person by means of writing or explaining. Visual elements provide the media and format for capturing and articulating "know-how".	Conventional logic-based training practices that fail to impart "know-how", or not offering on-the-job training.
Creating Shared Ownership: Visual media are utilized to remind employees of company vision, customer focus, and priority. Messages can convey support and encouragement. Employees understand the importance of their input at a glance.	Management dictating intangible vision and culture, and management pronouncement of change.
Management by Visual Facts: Fact-based visual information, free from personal bias, not affected by subjective interpretation. Employees trust is won over when there is an openness to share fact-based information especially when it is visually presented.	Management by unclear terms or subjective judgement. Absence of visual fact-based information can cause doubt of authenticity.
Simplification: Information is extracted from complex data sets to provide patterns, trends, anomalies and relationships to assist in identifying the areas of interest.	Expecting people to digest a lot of and/or complex information can lead to comprehension difficulties.
Unification: The use of visual means to frequently share information between related works of departments lowers barriers and creates empathy for internal supplier and internal customer needs.	People work on tasks according to their departmental values. "This is not my job" behaviour.

3.4 Attitude and mind-set affects actions people take

The development of a quality management culture starts when people begin to take actions to ensure the quality of their own work, and to ensure that things are "done right the first time". When things do not turn out right, the person should examine what led to the unsatisfactory output and act differently the next time in order to produce a different and correct result.

This seems so simple but is in fact difficult because people have to learn a different approach. This involves;
- their attitude when doing the work,
- their receptiveness to take different action,
- their understanding and ability to take different action,
- their will-power to persevere with and maintain different action

A person's attitude influences the action that they take. Changing attitude to influence a new action can be difficult, and can take a lot of effort and time. The person, be they a knowledge-worker or a manual-worker, needs to be receptive to taking a different action, and they must be adaptable to change. People are more likely to change their attitude when they expect a favourable outcome, and/or when they stand to win or lose something due to the issue.

Attitudes are likely to be held and strongly maintained if formed through extensive learning and personal experiences. People also tend to keep an attitude when it is repeatedly reinforced.

There are many people receptive to taking a different action and it is good to see them adapting well to change and reaping the success. However, there are people in every corner of this Earth who are inflexible or that resist doing things differently. Also, often found are people working nowhere near their full potential, but they are seemingly unprepared to commit to trying anything different.

The type of mind-set that the person adopts or acquires through conditioning, affects their notion to change, and affects their opinion about their abilities and the abilities of others. Mind-sets are defined by values and beliefs that set boundaries for new behaviour and give direction to new behaviour.

Carol Dweck[5], a researcher of why people succeed and how to foster success describes the "growth" mind-set and "fixed" mind-set.

She explains that those with a "growth" mind-set are able to use failure as a learning experience in order to improve over time. However, those with a "fixed" mind-set display resistance to change and hide or react defensively to mistakes or setbacks because deficiencies and mistakes imply a lack of talent or ability.

"In a fixed mind-set students believe their basic abilities, their intelligence, their talents, are just fixed traits. They have a certain amount and that's that, and then their goal becomes to look smart all the time and never look dumb. In a growth mind-set students understand that their talents and abilities can be developed through effort, good teaching and persistence. They don't necessarily think everyone's the same or anyone can be Einstein, but they believe everyone can get smarter if they work at it."

Some top-managers, both in the West and in China, eager to grow their organisations and to improve quality, often have a tendency to favour and promote certain people and to give special treatment to those with whom close relationships are shared. Unfortunately, with this tendency comes a lack of expectation for the growth of persons not in this group and persons below a certain level; these persons receive little encouragement and incentive to improve their work.

Enterprises under the control of leaders with a "fixed" mind-set struggle to progress with quality improvement and innovation, even worse are examples in the corporate world where the "fixed" mind-set has led to the demise of organisations.

Examining the case of the bankruptcy of the Enron Corporation[6], business analysts agree that Enron created a company-wide "fixed"

---

[5] Carol S. Dweck (born 1946), University of Stanford professor of psychology, author and lecturer.

mind-set culture, and with the "fixed" mind-set culture, its employees were not able to admit and correct their deficiencies – and a company that cannot self-correct cannot survive. Studies have shown that the ability of the "fixed" mind-set to rationalise itself can be so strong that the person never sees outside of it.

Those at the top of organisations with "fixed" mind-sets pose a problem to the fostering of an environment supporting continuous evaluation for correction and improvement, and therefore the healthy growth of the organisation. This is because this type of mind-set radiates down the ranks and affects the actions people take. People react defensively to mistakes and deficiencies because these imply a lack of talent or ability, and show resistance to change.

In order for a quality management culture to develop, it needs people, beginning with those in top management positions, with the "growth" mind-set; people that encourage others without prejudice, that do not resist change, that look upon problems as challenges to improve, and treat failure as a learning experience.

The power of the growth mind-set is perhaps best portrayed in Jack Welch. He was chairman and CEO of General Electric between 1981 and 2001. During his tenure at General Electric, the company's value rose 4,000%.

Jack Welch[7] hired according to what the person had made of their life, not according to the person's pedigree or class of university that they attended. He demonstrated a belief in people's capacity for growth by spending thousands of hours grooming and coaching employees on his executive team. He showed that a "fixed" mind-set can be encouraged to change into a "growth" mind-set through coaching and wise praise. Jack Welch set the culture of "we can all do better, we can

---

[6] At the time of its bankruptcy in December 2001, Enron Corporation employed approximately 20,000 staff, and was one of the world's major electricity, natural gas, communications, and pulp and paper companies.
[7] John Francis "Jack" Welch: (November 19, 1935) is a retired business executive, author and chemical engineer.

improve, and we can grow". With this attitude, a person's actions can be positively influenced.

Coaching a person with a "fixed" mind-set would address how they interpret challenges, and deal with criticism and overcome setbacks. Wise praise includes praise of the action strategy, focus, process, perseverance, effort and progress.

3.5 Influencing and encouraging behaviour

The behaviour of a person in the work situation has been studied and analysed over many decades to understand the factors influencing behaviour. A significant development occurred in the early 1970s when human performance was presented in a systems framework. This gave birth to Human Performance System (HPS) Models. These models are useful to those wanting to improve human performance and human behaviour in the work situation.

The principles behind these models are that every organisation has a system of work processes designed to transform inputs into valued outputs for customers; that every work performer, from the Chief Executive Officer to the production line operator, is part of a unique personal performance system; that when the work performer fails to produce a desired outcome, it is usually due to the failure of one or more of the components of that work performer's HPS, which is illustrated in figure 3-1.

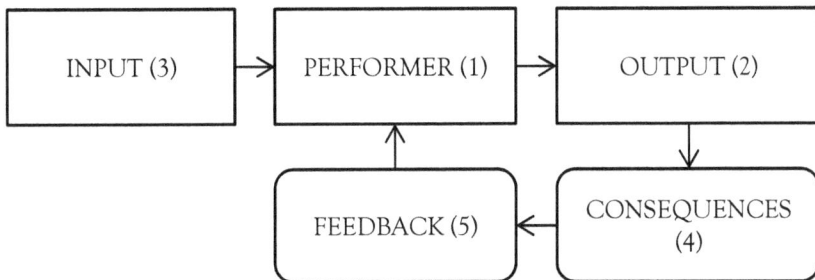

Figure 3-1: The Human Performance System

With reference to figure 3-1 the operation of a work performer's HPS can be summarized as follows:

- The work performer (1) produces outputs (2); for each output there is a set of inputs (3).
- For every output produced, as well as for the action it took to make the output, there is a resulting set of consequences (4). These consequences are the things that happen to the work performer which are interpreted by the work performer as either positive or negative. This interpretation is the key to understanding the work performer's future behaviour, because the HPS is governed by the behavioural law that people's behaviour is affected by consequences.
- Feedback (5) is given to the work performer (1) about the resulting consequences (4).

The reasons for sub-optimal and poor work performance can be due to, for example, a lack of meaningful feedback, no positive consequences for succeeding, lack of preparation, lack of training or no clear direction, shortage of materials, bad information, unavailable tools or equipment, and software issues.

There is also the impact of negative consequences for doing a good job that must be considered such as where an operator is required to mark-up defects from his work process but the piece-work scheme penalizes him, or where a person that persists in reporting a problem issue is told to stop being a nuisance.

Practical use of the HPS model:

The HPS model can be used for diagnosing human performance issues. Examination of the work situation against the conditions for an ideal HPS, as shown in figure 3-2, will help to identify issues that need to be addressed to improve and encourage human performance.

INPUT (3)	PERFORMER (1)	OUTPUT (2)
• Performance expectations are clear • Necessary resources for work are in place • Instructions are clear • Received components do not need fettling, clean-up or rework • No or negligible interference from extraneous demands	• Has the skill and capability to do the job (mental, physical, and emotional) • Knows where to get job-related information • Possesses a good understanding of why job is important • Is willing to perform, given the incentives available	• Appropriate criteria are known with which successful performance can be judged • Output is aligned with business success • The requirements of the output performance motivates possibility of further personal development

FEEDBACK (5)	CONSEQUENCES (4)
• Feedback is expressed in terms of job performance; it focuses on how well or how poorly the job is being performed • Feedback is frequent and relevant • Feedback is easy to understand and actionable	• Consequences and incentives encourage expected performance • Consequences are fair, timely, and consistently applied • There are no real negative consequences or disincentives to perform

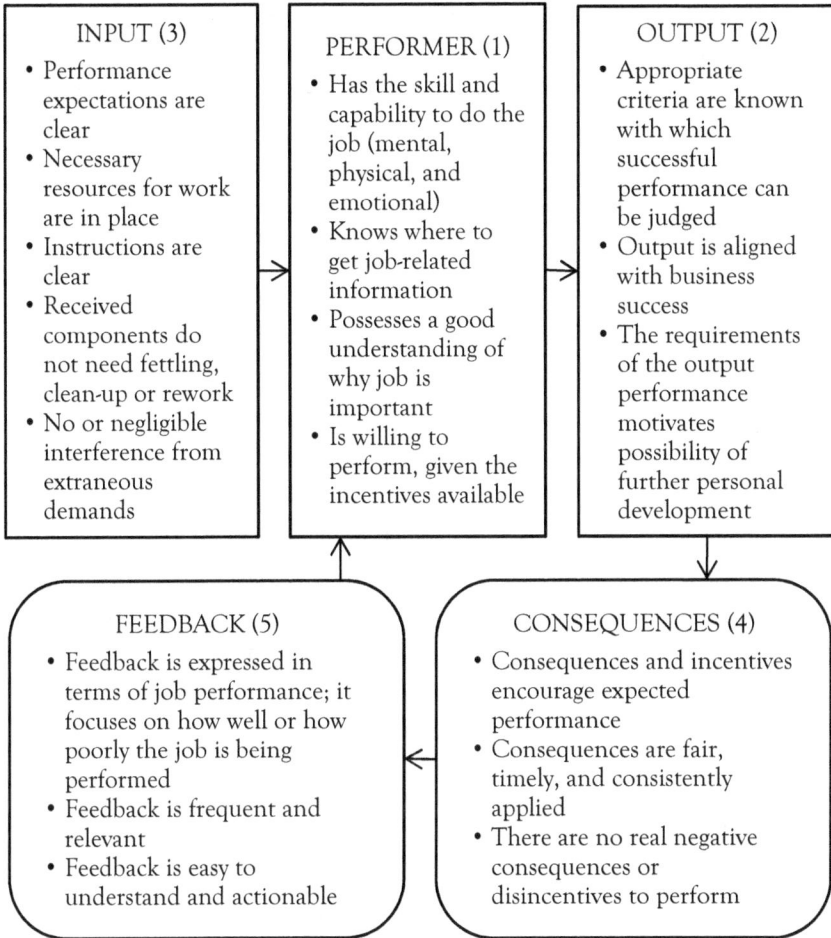

Figure 3-2: Description of ideal Human Performance System conditions

The following narrative serves to illustrate the use of the ideal HPS:

A manufacturing company had implemented Cellular Manufacturing as part of their World-Class Manufacturing initiative.

The pull-system through the plant meant that work-in-progress was kept to a bare minimum. The low work-in-progress allowed almost no tolerance for work stoppage; but many work stoppages and bottlenecks occurred due to defective product being released from the cells, as well as manpower problems.

There was a tendency for some of the male workers to over-indulge in alcohol over the weekend, with the result that some of them were not fit to work on the morning following the weekend. The company had a medical facility staffed by full-time qualified nurses and part-time medical doctors. The production supervisors would send the male workers with "weekend after sickness" to the medical facility, but, as we know, no medicine or medical treatment can instantly cure a person who has over-indulged in alcohol, and cure him to a state of being fit to concentrate on work, and work in a safe manner. Counselling and lectures had no impact. The company was at a loss for ideas of how to deal with this social problem.

The focus turned to investigating how to eliminate the many work stoppages and bottlenecks occurring due to defective product being released from the cells. The work situation was examined against the conditions for an ideal HPS described in figure 3-2.

Results of the examination showed the following:

PERFORMER (1):
Whereas the work performers had the basic skills to perform their jobs, some of them needed skills training to improve their level of proficiency. It was also established that the importance of the work performers' jobs had not been adequately stressed.

OUTPUT (2):
Although appropriate quality criteria had been included on technical documentation, the documentation was normally held by the cell leader and it was not always conveniently accessible to the work performer.
Operators needed constant reminding of the order or sequence of assembly, and what was important such as not mounting braided copper earth straps directly to the galvanized casing – it had not been explained to the operators that, especially in coastal installations of the

rail-side cabinets that they were building, corrosion would rapidly occur between copper in direct contact with galvanizing.

Furthermore, some criteria on technical documentation were not clear, e.g., the stipulation "wiring to be neat" did not convey what was required for wiring to be acceptably neat.

There were no job performance drivers in place to motivate personal development

INPUT (3):

The cell workers frequently received components from feeder cells that needed corrective action or cleaning-up.

The importance of the job performer's work in the contribution to the success of the company's continued operations was not understood by the workers.

The work performers operated almost in a vacuum, e.g., they were not informed of major happenings and significant achievements of the company – they relied on hearsay and rumours.

CONSEQUENCES (4):

The question was raised as to what incentives would influence them in a positive manner. The influences of bonus, ownership, pride and self-worth were discussed.

FEEDBACK (5):

It was evident that feedback was not frequently given to workers on how well or how poorly the job was being performed. Defects found on released product were often corrected by the recipient cell workers and, in many cases, feedback was not provided to the originating cell.

From these findings and realisations of the effect of certain shortcomings in the company's communications, the following actions were identified for each cell:

1) To hold a short meeting every morning with a set agenda. The meeting was to be no longer than 10 minutes. The agenda items:

(a) The cell leader would begin the meeting by reading out company announcements (these were called "management briefs") which included such things as winning a big order or doing or achieving something of great significance, or the visit of a customer (and purpose of the visit).

(b) The cell leader would then provide feedback on actions taken to resolve quality and operational problems reported the previous day.

(c) The cell leader would announce the daily target and invite discussion on foreseen obstacles in reaching the daily target, e.g., material, quality, and manpower issues.

2) To provide quality awareness training to all work performers.

The Quality Department arranged on-site quality awareness training for each cell which included the definition of quality, explanation of internal supplier/customer requirements, quality control actions in "Operator Self-Control", and a hands-on learning exercise.

3) To hold each cell responsible for the quality of their output.

This meant that non-conforming product would be returned to the cell responsible, and that the cell responsible would need to apply corrective action.

4) To design and implement a skills matrix, as shown in figure 3-3, to indicate skill-level status for each cell member.

Status indications would be as follows:

- "I": *Can do the job to the correct quality standards, first time, without assistance*
- "L": *Can do the job in the standard time, to the correct quality standards, first time, without assistance*
- "U": *Can do the job really well; recognised as an expert to train others in the job*
- "block is left blank": cell member unable to do that particular job.

5) To create a display board for each cell to display the skills matrix, quality charts, good production output, production targets, and quality control visual aids.

ILU SKILLS MATRIX CHART: SIGNALLING EQUIPMENT CASE											
I	Can do the job to correct quality standards, first time, without assistance										
L	Can do the job in the standard time, to correct quality standards, first time, without assistance										
U	Can do the job so well; recognised as an expert to train others in the job										
	Unable to do the job										
Work Performer		Banjies	Beste	Kloppers	Mampane	Mann	Morcomb	Moss	Peters	Smithers	Styles
Mechanical framework assembly		L	U	L	L	L	L	L	L	L	L
Inner mounting frame assembly		L	U	L	L	L	L	L	L	L	L
Electrical module installation		L	U	L	L	L	U	L	L	L	L
Internal wiring		L	U	L	L	I	U	L	L	L	L
Configuration set-up and performance test		I	U	I	L		L	I	L	L	I
Safety test and release inspection			U		L				L		

Figure 3-3: Example of a Skills Matrix Chart

The actions mentioned resulted in a much improved atmosphere in the cells and much better work flow. The previously occurring bottlenecks and work stoppages were eliminated. Defects occurring in the cell decreased significantly, and product released from one cell to another, or released to stores for dispatch to the customer, was found consistently practically defect free.

The short morning meetings served to unify the group to meet daily challenges; the cell members felt recognised and well informed, and knew what was expected of them in the grand scheme; they were particularly happy that quality and operational issues were openly discussed and actioned for resolution.

The skills matrix became known as the "ILU Skills Matrix Chart". The cell leader could tell at a glance which job performer was skilled at what job, and multi-skilled work performers could swap jobs with each other. Having a "U" against their name for as many cell jobs as possible became a symbol of achievement and peer recognition for job performers. An interesting benefit was that improvements in assembly methods were identified by multi-skilled job performers when they were moved to downstream positions.

Immediately after the morning meeting, the cell leaders would pass information (e.g., quality and operational problems needing resolution) to the production section leader who communicated this onwards for reporting and action purposes. Engineers and senior people frequently attended the 10 minute meeting to listen and contribute. The discipline was respected that discussion beyond the 10 minute limit was not allowed; if necessary, the cell leader would arrange another time for focused discussion with the person or persons involved.

The display boards carried the production targets, good production output, and quality performance charts showing type and number of defects (cumulative average against weekly) and the effect of corrective actions. The display boards also carried the "ILU Skills Matrix Chart", and quality control visual aids of photographs depicting acceptable and unacceptable electrical wiring, key point charts reminding workers of important criteria (e.g., earth strap assembly), and isometric drawings to indicate order or sequence of assembly. The cell workers were indeed proud of their display board; the display boards certainly attracted a lot of attention from management and visitors.

An unexpected development was that the number of job performers suffering from "weekend after sickness" dwindled away to zero. Perhaps this was due to fellow job performers taking pride and responsibility in the performance of their cell, and seeing themselves as a team where each member contribution was valuable and so putting these workers under some sort of cell-community pressure.

The actions arising from examination of the work situation against the conditions for an ideal HPS, such as those given in the narrative, gives an insight into what can be done in a more permanent way to influence behaviour of work performers. The consequences of their behaviour are all around them with constant reminders – they repeat behaviours that bring them positive consequences and avoid behaviours where reward is negative.

It appears that influencing and encouraging behaviour within industrial manufacturing organisations in China is accomplished primarily through a reward/punishment system, and recognition by way of trophies or achievement certificates.

A reward/punishment system may have its place in promoting compliance, and serve some purpose in crises and emergency situations, but reward/punishment has its shortcomings in that it does little to alter the attitudes that underlie behaviours, and it does nothing to create an enduring commitment to any quality management value.

3.6 Effects of communication

Expected behaviour is communicated in the organisation by internal communication, and internal communication helps to create shared values and beliefs. In the better examples of industrial manufacturing organisations in the West, the climate is set for an action-driven and quality focused management team through clear communication. Bureaucracy is avoided, and bureaucratic behaviour is regarded as unproductive and unacceptable.

The need for responsiveness leads to the tendency to keep organisation structures as flat as possible. In a flat organisation structure, top management does not have intermediary managers between them and their teams; they maintain direct contact with their teams and communicate with them directly on a regular basis.

The top managers in large manufacturing organisations in China are generally separated from their front-line managers by intermediary managers and therefore removed from the daily action. "Open-door" practice is rarely seen. Communication of information from subordinate managers to the top managers is frequently shaped and filtered. It is believed that cultural influences of saving face, protection of self, and maintenance of harmony cause this shaping and filtering of information.

This filtering and shaping of information also naturally occurs in organisations in the West because similar influences are at work, however, in the better organisations in the West, communication between top management and their subordinates has improved over the years to focus on the process and the fact-based technicalities of the issue, and actions that follow are directed towards the process and not the person.

A situation is recalled in an organisation in the West during an executive management reporting session in the late 1990s. At this meeting, the senior managers were required to report the monthly performance of their business or production units to the board of directors.

The Managing Director at the top of the management hierarchy was a surly man in his senior years. He wielded a great deal of power that was largely fear-based, and believed in the "carrot and stick" approach of offering rewards and punishment to induce the required behaviour.

The recently appointed senior manager in charge of the switchgear production unit delivered his report. The report followed the format of his predecessor with the addition of a presentation slide showing a Pareto Analysis of the various defects that occurred in his production unit in the month. The analysis indicated that most defects were due to burnt terminals on test.

The Managing Director interjected and told him in a loud, stern and threatening voice that these rejects were unacceptable; after all, he concluded, the problem was due to sloppy assembly operators not tightening terminals adequately and this was surely a simple issue to sort out. The Managing Director told the senior manager that he had better take immediate action.

When the senior manager returned to his production unit, he called all of the assembly-line operators together in the workshop. He sternly told them that they had better take immediate action to stop their sloppiness that was causing loose electrical connections. He then outlined to them a reward/penalty scheme based on inspection and test fault statistics – non-recoverable defects were to result in maximum penalty. The message communicated was clear!

At the following month's executive management reporting session, the senior manager in charge of the switchgear production unit presented his report. The chart of the various defects indicated no burnt terminals on test, and the next month the slide again indicated no burnt terminals on test. The Managing Director was delighted and repeatedly congratulated the senior manager on this great achievement.

The senior manager of the workshop believed that his stern and threatening talk and introduction of the incentive scheme had cured the problem because the information that he received from the assembly-line inspection and test area indicated no burnt terminals on test.

Six months later, the company's security reported finding numerous switchgear components on the adjacent, unused, industrial site, close to the company's perimeter security fence. The pile had grown so big in one area that the tall grass could no longer hide the mountain of components from the perimeter security cameras.

The assembly-line operators had responded out of fear. Any non-recoverable switchgear component was smuggled out of the workshop under the operators' work-uniform and thrown over the fence. The operators had been doing this for nearly three months. They had cleverly avoided being detected by the perimeter security cameras. The information given to the senior manager, concerning defects found during inspection and test, was filtered. Stock records were cleverly adjusted to hide the loss. The senior manager in charge of the production unit reinforced this behaviour by telling the supervisor and the operators how happy he was with their work.

The burnt terminal problem was analysed and found to be caused by connections not being adequately tightened. It was also noticed that the drawings and work instructions communicated no information regarding tightening torque of terminals carrying high electrical current.

Another finding was that loose connections were concentrated in a particular area which was difficult to reach in the fully assembled switchgear unit. The operators had communicated this difficulty to the production supervisor on many occasions, but the production supervisor had regarded the operator's communication as an excuse to cover up a lack of job expertise.

Once identified, the problem of loose terminations was readily eliminated; the sequence of build-assembly was altered to overcome the access issue, tightening torque was specified and appropriate tools provided for applying this torque. Infrared inspections were introduced at the heavy current test station to detect increased heat due to increase of resistance caused by loose and faulty connections.

The reward/penalty approach can communicate the wrong message. In the situation where the manager implemented a reward/penalty scheme based on inspection and test fault statistics, the manager thought the scheme was appropriate. Initial results outwardly indicated to him that the reward/penalty scheme worked well. The reality as seen by the operators was fear and loss, and, because they did not have the knowledge, training or ability to correct the problem, out of desperation and thinking that the manager was unfair, they behaved in a manner to protect themselves.

The reward/penalty or "carrot and stick" approach is very much prevalent in large manufacturing organisations in China. It is believed to give security of order and direction, and communicate what is important. But, many reactions have been witnessed that indicate that the approach is not as successful (at least in the management of quality) as some Chinese managers may think it is; to cite a few examples:
- workshops not declaring all of the scrap produced,
- operators not declaring defects caused by their work process,
- workshop personnel having bad feelings for Inspectors, resulting in poor co-operation

This approach needs to be re-thought and modified to ensure that a fair incentive is communicated.

Management by Walking Around (MBWA):

Powerful positive messages can be communicated by this management practice. MBWA[8] was personally first seen in action in the late-1970s at a plant designing and manufacturing automotive electrical components.

Almost every morning, the Managing Director of the plant would, after placing his briefcase in his office, immediately and without sitting down, begin a solo informal tour of office and work areas. He would pause and talk briefly to an employee; he would sometimes pick up a work-piece and examine it. He would comment on quality, how effective the employee was controlling the quality, or the excellent job the employee was doing taking care of the machine or workstation.

These tours of his were only 30 or 40 minutes in duration. The plant was large but he must have made a mental note of where he had been in a week and what areas he needed to re-visit; he seemed to appear regularly in practically every location in a plant employing 1,200 people – almost every area had their tale to tell of his visit. I remember this man visiting me, a young engineer, in the design and development laboratory. His visit was brief, but the impact was huge.

The overall result was high worker morale, a well-organised and neat workshop, and a low level of quality defects. This man showed that he cared and he "walked the talk".

Research suggests that the physical presence of the organisation's leader at the various places of work, and his attentiveness and interest in them and their work, shows the employees that he cares – and the effect invariably is that the employees show that they care about their work. In addition, the act of the leader doing MBWA gives employees the impression that the leader knows what is going on, and an understanding that the leader is taking the bigger picture in mind

---

[8] In the late 1970s, Tom Peters and Bob Waterman began an extensive research of companies that did things really well and achieved superior results. Among the best practices identified was MBWA. They published their findings in the best-selling book "In Search of Excellence". MBWA remains to this day an acknowledged best practice.

rather than blindly following carefully worded reports and biased opinions or beliefs of his first-line managers.

MBWA requires a relaxed and non-threatening type of engagement with employees to elicit open conversation, and many forces come into play which can challenge relaxed openness. However, where MBWA is practised, a positive atmosphere seems to always prevail, not only among office and factory workers, but also in the boardroom. The overall perception is "the boss is very interested and fully aware of what goes on". This leads to a healthier and honest communication where the boss does not receive filtered information, or is told half-truths.

Table 3-2 contains tried and tested guidelines for MBWA.

Table 3-2: Guidelines for MBWA

- It is better not to bring a group of assistants with you. Talk with each employee on a one-on-one basis and remember to lighten up and relax (share your hopes, bring good news about goals achieved and how their work helped); this will encourage more honest dialogue as well as contributions without intimidating contributors.
- Do it in every location of the organisation, as often as you can.
- Ask questions without being critical. Enquire how jobs can be improved as well as about workplace practices in general.
- Watch and listen. The goal should be to uncover honest and objective contributions, and not have employees feeling they need to tell you only what you want to hear.
- Do not confine your talks only to business matters.
- If you find that an employee is not performing his or her job correctly, do not criticize or attempt to change behaviour on the spot. Make a mental note and address the situation with the employees' supervisor at another time and in another setting.
- Reward valuable contributions in some way. This does not have to be a monetary reward, some type of certificate or company memorabilia will demonstrate that you recognise good ideas.

My first and possibly most powerful experiences of the power of MBWA in China were through regular tours with the General Manager of a factory situated in northern China.

The General Manager, a charismatic and well respected senior engineer, would casually walk to the various workstations in the factory, examine the quality of the product, ask about process control measures being taken by operators, and enquire after quality controls being used. The workers saw that he was concerned about the quality of the product, and that he was interested in their work.

The General Manager was also present at quality and process training and education sessions where his comments and questions reflected his keen interest.

A practical example of the power of his MBWA tours is reflected in improvements that followed shortly after he had attended awareness education sessions concerning Operator Self-Control. The sessions covered the necessary prerequisites required to make Operator Self-Control successful and, during his MBWA tours of the factory, the General Manager began looking for evidence that appropriate means were in place.

He asked various employees informally about Operator Self-Control and checked, for instance, that the work environment was satisfactory, that the necessary means to perform self-checking had been provided, and that things to check had been made clear. He asked what actions the operator would take on finding problems of various types and checked that a system of identifying and segregating non-conforming product was being used. He had learnt that the Process Engineering Department had put a lot of effort into developing a new generation of process procedures and sought confirmation of their use and of their usefulness to the operators.

These observations and questions uncovered significant issues, for instance:

- Although in-process measuring instruments were available for the operators to use, these instruments were kept in locked cupboards, and these cupboards were a fair distance from the operators' workstations; some cupboards were as far as 100 metres distant.
- The excellent process procedures were not available in the production workshop in any form for operators to use; their use was confined to more senior personnel who were tasked with job set-up.
- Some operators that performed critical-to-quality work were unaware of the high value of their work to the customer.

The General Manager followed up his findings which resulted in changes such as: at job set-up, together with the issuing to operators of jigs and tools needed for the job, quality check gauges were also issued; in addition at job set-up, operators were briefed regarding critical-to-quality characteristics, and quality control information was issued to operators. The quality control information was extracted from the process procedures and placed on charts that were required to be displayed at the workstations.

The regularity of his MBWA tours resulted in a general sharpening up on efforts made by practically all involved in one way or another.

Over a three year period, the gains brought about by many important changes both big and small, as well as innovative processes and the implementation of effective quality management, resulted in a four-fold increase in the profitability of his factory.

Questions for Chapter 3

3-1: Describe some important actions that managers and leaders can take to positively influence a company's quality management culture.

3-2: Give typical examples of how Visual Management in a manufacturing organisation is used to help control and achieve quality requirements and influence a desirable quality management culture.

3-3: Summarize in key points your understanding of the Human Performance System model.

3-4: Describe the conditions for an ideal Human Performance System.

3-5: Give examples of internal communication of senior management that supports and encourages the desired behaviour of a quality management culture, and examples of internal communication that oppose the desired behaviour. Include the likely reaction in the company to these different styles of internal communication.

PART 2: OWNERSHIP & MANAGEMENT RESPONSIBILITY

Subjects, practices, concepts and techniques contained in Part 2

Chapter 4: Quality ownership
- ♦ Ownership of action
- ♦ Response in a closed loop control system
- ♦ Operator Self-Control
- ♦ Ownership of quality management processes
- ♦ Quality department activities
- ♦ Typical responsibilities of the Quality Manager
- ♦ Typical work of the Quality Engineer

Chapter 5: Management's responsibility
- ♦ Emphasis of top management to take the lead in quality management
- ♦ Understanding and meeting customer requirements (and the job of contract review and technical review)
- ♦ Strategic planning and the inclusion of quality goals and objectives
- ♦ Balanced set of quality objectives
- ♦ Deployment of goals and objectives
- ♦ Aligning Key Performance Indicators with goals and objectives
- ♦ Ensuring clear and effective communication processes
- ♦ Use of information for quality improvement

4. QUALITY OWNERSHIP

The quality of a product or service provided by an organisation is the result of the decisions, behaviour, and work of every person within that organisation. Everyone owns a responsibility in the making of a quality product or service.

This chapter discusses:
1. Ownership of actions
2. Ownership of the quality management processes
3. Activities of the quality management function

4.1 Ownership of actions

In the tasks we do in daily life we naturally control our actions using feedback. To illustrate this, in the action of cutting a piece of shelving with a hand-saw, having measured and drawn a line, we guide the hand-saw to follow the line; when we see our cut moving away from the line, we observe this deviation and use feedback to correct the deviation by applying a force in the direction on the handle of the saw to bring the cut back to the line. There is perfect co-ordination between our actions. Our actions describe a closed-loop control system.

A system is a combination of components that act together. When we own all the components of a work system – the doing, the observing, and using feedback of the observation result to control action to achieve the required output – we quickly learn to do a good job.

The closed-loop control system is naturally used in the many tasks being performed in an organisation. Figure 4-1 illustrates the operation of a closed-loop control system which is further clarified as follows:

Before we decide to cut the length of shelving, we would have decided what the required measurement of the length must be (Output, 2), and we would use this measurement value to determine where to draw the line for our cut (1), providing us with a reference.

The "Dynamic Work Unit" would begin the operation – in this illustration this would be us using the handsaw. (The "Dynamic Work Unit" could be practically any type of work operation.)

Our cut may wander slightly off the line because the blade in our hand-saw had not been tensioned correctly, or we may be inadvertently applying sideways pressure to our hand-saw; this is shown as "Disturbance" in figure 4-1 and arises from "special cause" issues. (Refer to Chapter 10.1)

We will use our observation as feedback (4) to compare how our cut is aligning with the reference line that we drew on the piece of shelving (1) and re-tension the hand-saw blade or correct the action of our sideways pressure (3). In this way the action of the "Dynamic Work Unit" is guided to achieve the required output (2), i.e., the controlled variable.

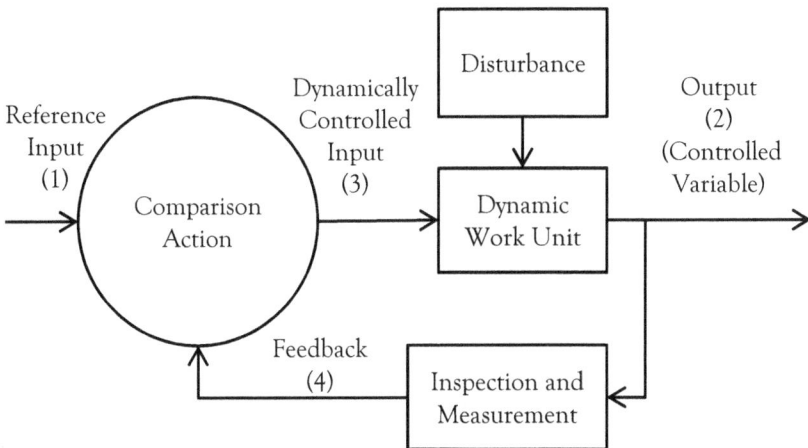

- The Input (1) value is set in terms of the required value of the Output (2).
- Disturbance to the Dynamic Work Unit arises from unwanted elements that affect the Output (2).
- Inspection and Measurement of the value of the Output (2) is performed, and Feedback (4) concerning variability of the value is obtained.
- The Dynamically Controlled Input (3) value is obtained from comparison of Reference Input (1) value and Feedback (4); this is applied to the Dynamic Work Unit to maintain the desired value of the Output (2).

Figure 4-1: Functional block diagram of a closed-loop control system

This seems straightforward, however, encountered numerous times in industrial manufacturing organisations in China, the feedback (4) has been absent or delayed, therefore immediate dynamic control (3) of the work could not happen. This was particularly the case when the ownership of the different components of the system were separated; the most common was when inspection and measurement was performed, or assumed to be performed by another party, or was considered unnecessary. In these cases, the work was not in an effective closed-loop quality control system and the persons doing the work were not using feedback to focus themselves on controlling the work to maintain quality conformance.

In the typical arrangement in many industrial organisations in China, this separation of the different components of the system occurs where the Manufacturing Branch is responsible for producing the products and Central Inspection are responsible for checking the quality conformance of the products.

If inspection is done by another party, time to feedback is critical if a closed-loop control system is to be preserved. The longer the time to feedback, the less "dynamic" the control becomes. Feedback occurring hours after production is of no real control use. A closed-loop quality control system only has a chance of working when the components are inspected as they are produced. Of course the immediacy in timing between production and inspection could be affected by the nature of the characteristic performance of the "Dynamic Work Unit" on the output. Once this is known, inspection may be considered to be done on a less frequent or sample basis. Trained Inspectors can work in this manner; they understand variability and they understand their responsibility in providing feedback on variability in order for dynamic quality control to occur.

The better arrangement of closed-loop control is through "Operator Self-Control". With this there is perfect co-ordination of actions because the Operator does the in-process inspection and makes adjustment as required to maintain control.

In certain large manufacturing facilities in the West it is usual to find Patrol Inspectors monitoring and supporting "Operator Self-Control" at each workstation, as well as monitoring the overall quality control in the workshop.

The Patrol Inspector will, for instance, perform random checks at workstations on components in the various stages of production to ensure that their quality status is clearly identified and that they conform to quality requirements. The Patrol Inspector will also assist with the prompt resolution of quality issues that may arise from workstation supplier and downstream processes.

In brief summary,

- A personalized view of our actions to reach a required output is described by the closed-loop control system.
- When the requirements of our output are known and absolutely clear, a basis for quality achievement is set.
- When we own the components of doing, observing/inspecting, and correcting through the use of feedback to control, we have perfect harmony, co-ordination, and timing between our actions.
- When we do not own all the components, we had better make sure that there is excellent harmony, co-ordination, and timing between the components if we are to maintain our effective closed-loop quality control system.

4.2 Ownership of the quality management processes

Up until the 1980s in manufacturing organisations in the West, a Quality Manager was given the responsibility of implementing the quality management system. He arranged for the writing of quality procedures according to his understanding of what should be done to meet ISO 9001 requirements. This was far from satisfactory.

The reality is that the Quality Manager and the quality management function cannot "implement the quality management system". The act of implementation can only be done by the people that do the work, i.e., the owners of the various work processes. And the quality management system has many processes, and each has an owner, and the quality management function is also an owner of a few of these processes. For example, the quality management function would most likely be responsible for organising internal quality audits, and organising or performing quality awareness education in quality.

In progressive Western organisations the acceptance of ownership of quality management processes began when top management made it clear to employees that they, top management, fully supported the necessity of every quality management process. Management persevered with insisting that quality management processes were to be followed as part of normal work routines, and they arranged that education and guidance was given in order that employees comprehended the intention and obtained the correct outcome of each quality management process. When intent was understood, the quality management process in question became meaningful and therefore its acceptance as an effective way of working was ultimately achieved.

In late 1978, China embarked on the long journey of transforming its economy from centrally-planned to market-oriented. The "reform and open-door" policy heralded in a change in economic development strategy; it became possible to rent land on a long-term basis, permission was given to entrepreneurs to start businesses, price controls were removed from the bulk of goods and services, foreign investment was allowed, and the government began promoting quality management.

Just before the end of this century, official decisions were taken to reduce the weight of the state sector on the economy. State-run companies were scrutinized for profitability and many were privatised.

Opening up to foreign investment has continued and GB/T 19001 (ISO 9001) certificates are now internationally recognised.

The economic reforms and the opening up of China to foreign business interests brought about the introduction and transfer of a new way of business thinking based mainly on Western management knowhow. However, it has taken a long time to create a climate of acceptance of this management knowhow with implementation going through what may be described as an adaptive phase – quality management process ownership requires in general a lot more nurturing.

Perhaps the most glaring evidence of this is when top management are unable to explain their responsibility in quality management, or the documented quality management system presents a reasonable picture in the boardroom but visits to work-places reveal a different and less flattering picture; deficiencies in the basic understanding and implementation of quality management emerge, and there is an obvious absence of ownership of quality management processes.

Such evidences frequently arose during companywide quality management system audits of manufacturing organisations claiming compliance to GB/T 19001 (ISO 9001). Certification appeared to have been obtained without proper qualification, and procured primarily to look good to the market in response to the increased competition.

Given the cultural dynamics in Chinese organisations, top management play a pivotal part in getting the various functions of the organisation to accept ownership of quality management processes. It is believed that education and guidance would bring huge benefits, but this needs to begin at the top of the organisation.

A good response to teachings in various quality management processes was experienced and it was noted that when intent was understood, the manager of the Unit would cleverly incorporate the quality management process into the work routines of the Unit through various means such as forms, computerized workflows and check sheets.

4.3 Activities of the quality management function

Activities of the quality management function or department have developed with the purpose of keeping the organisation's performance in line with intentions, and of helping to achieve the organisation's goals of customer satisfaction and company profitability.

Nowadays, quality action activities are as prominent in the design and development arena as they are in the production arena. This is because the quality of the manufactured product arises not only from the "quality of conformance", but also from the "quality of design". The "quality of design" is also referred to as the "quality standard"; this is the quality specified by the designer in drawings and technical specifications. (It can be said that the quality is actually specified by the designer on behalf of the customer because the product is designed to satisfy customer requirements.)

In the design and development arena, reviews, verifications and validations need to be carried out because the "quality of design" can be influenced by such things as skills of the designer, the actual concept of the solution, materials chosen, technical processes specified, dimensions and tolerances selected, and software programs used.

On the other hand, the "quality of conformance" in the production arena can be influenced by the skills of the operator, variance in materials, variability of technical processes, manufacturing method chosen, the tools selected for the job, the variability of the machines selected, and the measurement and feedback process.

The Quality Manager in a large manufacturing organisation in a Western company usually reports directly to the Managing Director and is part of the executive team. In this position he is able to apply systems thinking (see Appendix 1) and ensure that the quality management system is effective in all areas and functions of the organisation, and he can give advice and guidance on leading quality matters. Thus, typically the responsibilities of a Quality Manager in the West move beyond "ensuring the establishment of an effective quality management system"; these responsibilities include the following:

- Identifying factors relating to quality which are important to the success of the organisation, through understanding and analysing the business statistics and customers' needs and requirements, and risk of failure to satisfy customer requirements. He is therefore involved in the organisation's strategic planning process, and in reviewing existing policies and practices, developing quality goals and action plans for improvement, then guiding this improvement, and reviewing and assessing the effectiveness of changes made.

- Providing guidance and advice on quality management system implementation through total involvement at management level (as explained above).
- Ensuring that sales and marketing, and other functions in the organisation, are aware and take into account the quality and reliability requirements of clients, and that appropriate steps are in place for meeting these requirements.
- Monitoring how the quality management system is performing, and that all managers, process owners and supervisors develop and maintain their part of the quality management system.
- Stimulating (championing, supporting or leading) continuous improvement (e.g., using the PDCA approach), and engendering a quality culture.

The work of the Quality Management Department is such that its team needs to be complemented by quality engineers with appropriate education and experience. Quality engineers are active throughout the organisation, from order enquiry through to field service; they are present in design and development, manufacturing, installation, and acceptance testing. Their work includes:

- Taking customer requirements and translating these into functional requirements.
- Establishing quality guidelines.
- Managing and participating in design reviews and bringing together key interested parties and subject-matter experts.
- Making suggestions to improve "quality of design" to make products more serviceable and manufacturable.
- Defining product inspection and test strategy and identifying specific inspections and tests required to be performed at various stages of product realisation.
- Initiating and driving the development of test processes and procedures.
- Identifying measurement and test tools and equipment.
- Performing tests to validate performance against design specifications.

- Identifying and championing continual improvement ideas in products and processes

Quality Managers in industrial manufacturing organisations in China do not appear to have prominence in influencing business decisions. In general, quality management functions (departments or offices) do not appear to serve the same purpose as they do in industrial manufacturing organisations in the West.

In class-leading manufacturing organisations in China, the Quality Management Office was found to play no part in defining the processes which could result in the production of quality products and services, and the prevention of PONC. There were no quality engineers or technically competent personnel from the quality management function assigned to affect control, evaluate and improve products and processes in areas of design, manufacturing, installation, and servicing.

The Quality Management Office performed duties such as;

- compiling quality procedure documentation,
- organising safety compliance testing,
- preparing for and arranging the annual internal quality audit,
- arranging the meeting of relevant people to resolve internal and supplier-related quality problems,
- liaising with customer's inspectors,
- reporting of statistics from quality improvement projects.

The "quality improvement" projects seemed to mostly favour the final product in the production domain; PDCA was not employed; the majority of serious and frequently occurring problems were not analysed to determine the root cause in order for corrective or remedial action to be taken.

These Quality Management Offices simply lacked personnel with technical strengths. Duties performed were not prevention-centred nor were they satisfactorily focused on continuous improvement.

There is quite a difference in duties performed when viewed against the typical duties of a Quality Management Department found in the better manufacturing organisations in the West; typically these duties are, in summary, as follows:

- Ensuring that customer needs and requirements have been accurately identified and that the organisation is consistently working at meeting these needs and requirements.
- Ensuring that quality standards, processes, policies, checks and product inspection and test plans are in place.
- Organising regular (e.g., bi-weekly) internal quality audits to determine the continual effectiveness of implementation of quality management processes.
- Verifying that applicable technical standards are interpreted and applied correctly.
- Performing or arranging validation and product proving tests.
- Holding or participating in design reviews to ensure that relevant design milestone criteria have been reached.
- Identifying quality verification measurement, inspection and test tools.
- Providing or arranging training and education to enhance quality control at the workplace, and in the use of quality measurement techniques and quality tools such as Check Sheets, Pareto Charts and Control Charts.
- Collecting and analysing data for identification of trends and opportunities for improvements, and encouraging others to do the same.
- Playing a major role in quality problem solving and encouraging continuous improvement.
- Supporting the development and maintenance of customer focus within the organisation.

The Quality Management Department in the better Western organisations also usually has a major responsibility in Supplier Quality Assurance which makes Lean Manufacturing and "Just-In-Time" production possible.

"Just-In-Time" requires that the organisation is absolutely assured that their suppliers are supplying them exactly what is needed, exactly when needed, so that the organisation does not have to stop to test and inspect each incoming item for defects.

In industrial manufacturing organisations in China, strong Inspection Departments and weak Quality Management Offices were found. The job of the Inspection Department was to inspect, and only inspect. The Quality Management Office generally held a low status in the organisation and a relatively low importance was attached to their work; quality audit findings were often brushed off by the audited parties, and response to corrective action requests were poor. This was believed to occur because of non-supportive organisational cultural values.

A management-led supportive organisational culture (refer to Chapter 2), together with development of the quality management function and of quality engineers, would bring big benefits to industrial manufacturing organisations in China. Efforts should be concentrated on customer satisfaction, identification of risks that could impede customer satisfaction, prevention of quality problems, and continuous improvement.

Questions for Chapter 4

4-1: What are the necessary conditions for an effective closed-loop quality control system?

4-2: Discuss the basic understanding required for effective quality management system implementation in an organisation, and actions of top management that are necessary for acceptance of ownership of the various quality management processes.

4-3: Where should the efforts of a developed Quality Management Department be concentrated to help an industrial manufacturing organisation achieve its goals of customer satisfaction and company profitability, and therefore what are the typical duties of a developed Quality Management Department?

5. MANAGEMENT'S RESPONSIBILITY

This chapter examines the responsibility of top management with regard to the leadership of quality management, under the following headings:

1. Changing emphasis of top management
2. Understanding and meeting customer requirements
3. Strategic planning and quality goals and objectives
4. Internal communications for effective quality management
5. Use of information to facilitate quality improvement

5.1 Changing emphasis of top management

ISO 9001 (GB/T 19001), the International Standard for Quality Management System Requirements, stipulates requirements for top management that were formerly contained in the standard under "Management Responsibility". The 2015 revision of ISO 9001 replaces "Management Responsibility" with "Leadership", emphasising that top management have responsibility for the leadership of the quality management system.

This emphasis in the responsibility of top management has evolved most notably over the past two and a half decades. Back then top managers in many Western organisations took "a courteous but shallow responsibility" in quality management: quality intentions were proclaimed, slogans were written and communicated by means of posters on walls, responsibility for quality was passed down the organisation and a Quality Management Office was required to set up the "quality documentation". Members of management obligingly sat through the management review and the external auditor meetings and learnt to say the right things to satisfy the objectives of the meetings.

Considering how management in organisations in China currently shows their responsibility in quality management, a wide range is evident from that described above, i.e., from "a courteous but shallow responsibility" to an active responsibility. The improvement towards an active responsibility appears in managers that have been at the receiving end of their customers growing quality demands, and in managers that have obtained a high level of quality awareness through national and international business interests, transactions, and factory tours.

In the 1990s, this shallow management approach was discussed by ISO quality management committees. These discussions brought major changes to the 2000 revision of ISO 9001 that were carried through to all latter revisions with clarifying refinements to emphasize the following:

- Top management have responsibility for the leadership of the quality management system and effective quality management requires their involvement and commitment.
- Top management must ensure that customer requirements are understood and met with the goal of improving customer satisfaction.
- The quality policy must identify the main goals of the quality management system and create a background for establishing quality objectives.
- Quality objectives, that support the quality policy, must be measurable and communicated throughout the organisation.
- Top management must set up an effective system of communication to ensure effective operation of the quality management system.
- Top management must regularly review the quality management system to make sure that the goals are being achieved, and to look for ways to improve its suitability, adequacy and effectiveness. The review must include assessing opportunities for improvement, and the need for changes to quality policy and objectives and quality management system.

The business environment has changed dramatically since the major revision; technology has changed how we work, distances have been dramatically reduced by increasingly efficient transport systems, the volume of international trade has increased, and electronic media (communications, email, the electronic office, internet and television) have made the world virtually a global village. ISO 9001:2015 takes into account the dynamics of the changing business environment. It enhances the previously stated requirements and requires that top management take a more active role:

- With "Leadership" comes a responsibility for top management to take "accountability" for the effectiveness of the quality management system.

- Requirements for leadership and accountability include ensuring that quality policy and objectives are compatible with strategic direction, and that quality policy is applied – not just communicated and understood.

- Top management are to ensure that quality management system requirements are integrated into the organisation's routine business operations, with risk-based thinking being built it into the whole management system.

The quality management system cannot work in isolation; quality management system requirements must be incorporated into the organisation's business operations.

5.2 Understanding and meeting customer requirements

It is the responsibility of top management to ensure that the organisation has ways and means in place for ensuring that customer requirements are fully understood and met to achieve customer satisfaction.

Responsibilities pertaining to meeting of customer requirements include:

- Giving advice and guidance for the best solution to the customer's needs, before sale, during sale discussion and after sale.

- Ensuring conformity of product to the agreed technical performance requirement.

- Ensuring the inclusion of technical requirements relating to safety, reliability, maintainability, and interchangeability to minimize equipment downtime.

- Paying attention to all the aspects that may not be stated in the contract such as statutory and legal requirements.
- Meeting the agreed delivery date.
- Giving clear installation and set-up instructions.
- Giving clear operation and preventive maintenance instructions.
- Providing prompt and professional after-sales service.

The processes of a well-thought-out quality management system serve to provide these ways and means for ensuring that customer requirements are fully understood and met to achieve customer satisfaction, and an important quality management process is that of contract review.

In industrial manufacturing organisations in the West it is common practice to have a contract review upon receipt of an enquiry to supply, or upon a request for quotation, and prior to the organisation submitting quotes and tenders, accepting contracts or orders, and accepting changes to contracts or orders. In this review, which includes a technical review, the customer's request is assessed with respect to whether the organisation can satisfactorily provide the solution and meet the technical requirements, and whether the organisation has the ability to produce the product, and produce it within the requested delivery time.

The intent of contract review is to ensure that the organisation understands and can meet all requirements before promising the customer that the product or service can be provided.

In manufacturing organisations in China with a GB/T 19001 quality management system, contract review type processes were found, but typically the contract review process was entered after the order enquiry phase and covered primarily the payment and legal conditions, and excluded technical review and an accurate check on the likelihood of the organisation achieving the customer required delivery schedule.

Technical review did occur but only after order acceptance.

Especially regarding the engineering specification of sophisticated industrial machinery, the vulnerabilities concerning the equipment's intended practical application and operating environment must be known before contract agreement; this is in order for the appropriate solution to be specified, otherwise the customer's expectations may not be met.

Often product failure is unjustly attributed to the manufacturer's production personnel and processes when in fact the product, as the solution to the customer requirement, has been sub-optimally specified. This indicates an inadequacy of the application of technical review that should have been carried out prior to order acceptance. The consequence could result in premature equipment failure: to cite an instance, in a failure investigation study in China involving fifteen installations of armoured face conveyor equipment, the load factor of conveyor chain, expressed as a ratio of estimated maximum tension from armoured face conveyor drive to declared operating force of the chain, was found to have a huge range. At least 50% of the installations had uncomfortably high load factors. In these installations, the stress placed on the chain in operation was so high that the chain was rendered highly susceptible to stress fatigue failure. In such cases of equipment failure, it was always the equipment supplier that was the loser in terms of money and often image.

The failure to take account of the intended operating environment could result in a number of product service life operational problems and shortcomings. For instance, not appreciating the effect that environmental conditions have on the product (e.g., acidity of mine-water, chloride in mine-water), not appreciating the effect on product reliability that operating practices have on the product (e.g., repeated shock-loading), and not understanding the consequence of negligence in preventive maintenance – in Chapter 10.4, capability of equipment in its intended application and the probability of failure are discussed.

The shortcomings in processes such as the contract review and technical review should have been detected through capable persons performing internal quality audits; then again, failure of internal quality audits to identify these shortcomings, and to initiate corrective and improvement action, could result in such issues going undetected for years and years.

Top management should, at the very least, ensure that evidence is presented to them during regular management reviews that the goals of the quality management system are being achieved, but, from observation, management review reports in organisations in China generally lacked this substance, and the contents often gave the impression that the report was generated primarily to fulfil the requirements for GB/T 19001 accreditation.

Top management needs to continually remind themselves, and appropriate personnel in the organisation, to look for ways to improve the suitability, adequacy and effectiveness of quality management processes. It is interesting to note that in ISO 9001:2015, top management will be required to be actively involved in the "operation" of the quality management system.

5.3 Strategic planning and quality goals and objectives

Top management in prominent manufacturing organisations in the West engage in strategic planning. This defines a clear mission and vision for the organisation, assesses both the internal and external situation and identifies issues critical to performance, and performs analyses to identify key issues from which an action plan emerges for the achievement of goals and objectives that lead to the vision.

Figure 5-1 clarifies some terms used in strategic planning; the strategy is the means by which the organisation plans to reach its goals and objectives that lead to the vision, and planned actions are required to achieve goals and objectives.

Quality goals and objectives must be included in the organisation's goals and objectives, and the quality policy – which shows the intent of top management with regard to the management of quality – serves as a guide in the strategic planning process and when formulating the action plans: the quality policy expresses top management's aims and commitment to meet customer requirements, and also to continually improve the effectiveness of the organisation's quality management system.

NOT EASILY CHANGED

Mission: The purpose of the organisation; its reason for existing.
E.g.: *To establish an interplanetary transport system.*

GETS ACHIEVED (typically annually re-visited; details can change)

Vision: Defines what the organisation wants to achieve in the mid or long-term future; provides guidance and inspiration.
E.g.: *Launch an interplanetary transport system with the first cargo vessel reaching Mars in 2027.*

Note: A Quality Policy is determined which serves as a guide in the strategic planning process and when formulating the planned actions.

Output of strategic planning:
Planned actions to reach goals and objectives.

Goals are set at various stages of the strategy and give the direction to move toward the Vision:
G1: *Build a fully operational Earth orbiting Refuel and Cargo Station that can meet the needs of the Mars cargo vessel, Excellence Evolution.*
G2: .....

Objectives are specific, measurable, with a defined completion date:
O1: *Complete construction, commissioning, inspection testing, and have in Earth orbit, a Refuel and Cargo Station (RCS) by 2024-09-30.*
O2: *Set-up a qualified supply chain to RCS with capability of meeting Excellence Evolution's needs; scheduled Mars launch is 2026-05-28.*
O3: .....

Mars

Earth

RCS

NOT EASILY CHANGED
Values, beliefs and behavioural guidelines of the organisation.

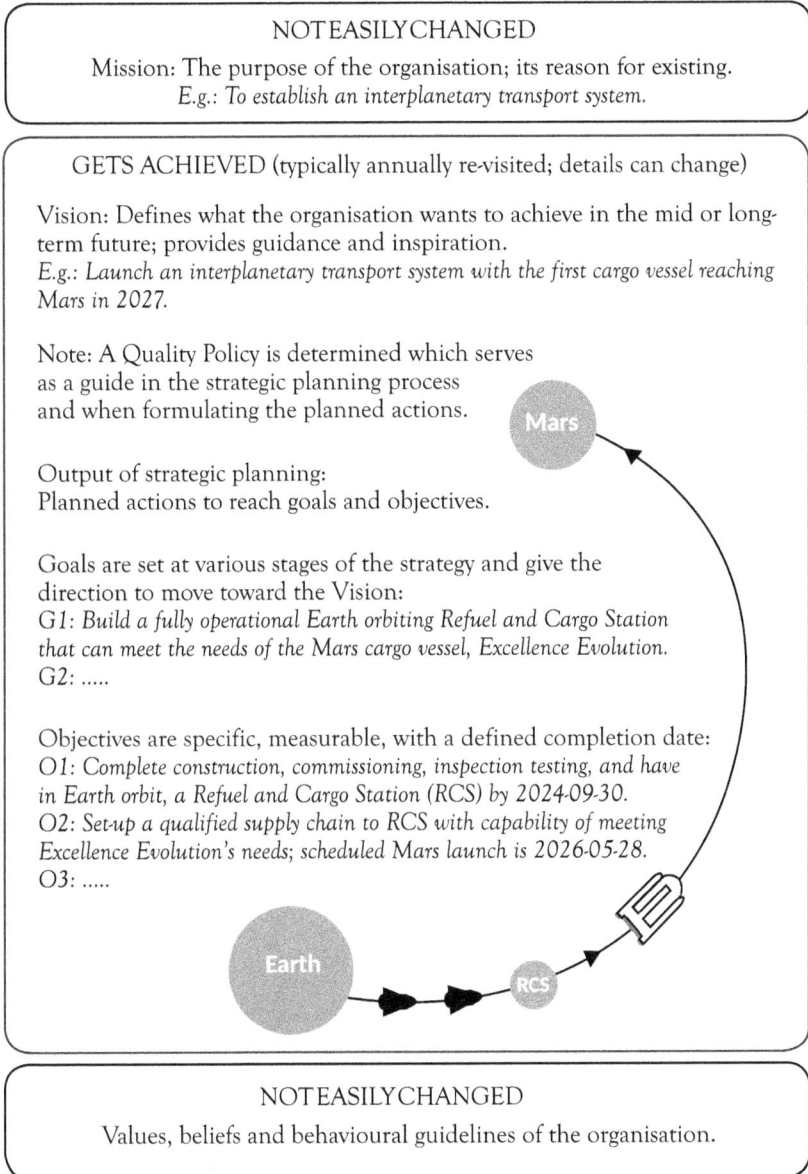

Figure 5-1: Strategy – Map of goals and objectives that lead to the vision

Table 5-1 contains the outline of a strategic planning process. This particular approach assumes fairly accurate forecasting and is therefore better suited for stable environments.

Table 5-1: Outline of a strategic planning process

1. Define the organisation's Mission (This should be previously defined)
2. Define the organisation's Vision
3. Perform an Environmental Analysis to identify External Issues
4. Identify Internal Issues critical to performance
5. Perform a SWOT Analysis
6. From the SWOT Analysis, identify Key Issues
7. For each key issue, identify actions
8. Assign actions and agree an Action Plan
9. Control the implementation of the Action Plan

Referring to table 5-1;

- "External Issues" are identified from an analysis of the macro- and micro-environment. The macro-environment – comprising political, economic, social and technological factors – affects all organisations. The micro-environment – comprising factors of substitute products, suppliers, niche customers, competing organisations, and barriers to entry – affects only those organisations in a particular industry.

- "Internal Issues" come from the values of the organisation, the image and reputation of the organisation, the quality and reliability of its products, position on the experience curve, key personnel, patents and trade secrets, exclusive contracts, operational capacity and efficiency, financial resources and geographical location (i.e., access to transport and utility services).

- "SWOT Analysis" – analysis of Strengths, Weaknesses, Opportunities and Threats – will draw out the "Key Issues". This directs attention to the strategy which needs policies and actions for implementation. "Key Issues" include factors that are critical to the organisation's performance; these could be: number of new orders, cash collection efficiency, return on investment (ROI), and factors directly pertaining to quality such as product conformity, customer satisfaction, employee competence and training, supplier performance, and overall quality system effectiveness.

- The strategic planning process will conclude with an Action Plan to reach the organisation's goals and objectives.

  An action arising from the strategic planning process could be for top management to review the organisation's quality management system. This may be necessary due to changes that emerge from strategic planning as the quality management system must be aligned to help the organisation achieve its planned goals and objectives.

Through the process described, meaningful goals and objectives can be identified that support the organisation's mission and vision, and an action plan agreed and followed to achieve the goals and objectives.

Balanced set of quality objectives:

Quality objectives need to be determined for a range of criteria that set an all-encompassing drive for quality in an organisation. The focus of the best manufacturing organisations in the West is not solely on product conformity, which seems to be the favoured focus of manufacturing organisations in China, but rather on a balanced set of objectives which include the following:
- customer satisfaction (performance against on-time delivery target, effectiveness in handling commercial and technical complaints),
- continual improvement (number of corrective, preventive and improvement actions, and effectiveness of actions taken),
- skills and competencies (a target number of reviews performed and actions taken to improve skills and competencies),
- supplier rating (number of suppliers reviewed and rated, e.g., per quarter, and improvement actions in supplier performance),
- and, product conformity (reduction of number of defectives, reduction of repeat problems, reduction of rework and scrap)

Deployment of goals and objectives:

Top management in the better organisations in the West ensures that the planned actions, or derivatives of these actions which relate to the work performed by each function and relevant level in the organisation, are discussed and agreed with these functions and levels. These actions have goals and, where necessary, performance objectives, and are planned such that achievements at each function and level contribute to the big picture. Figure 5-2 shows an example where the overall goal is to reduce the number of defectives in Production.

Production Director Goal: Reduce the number of defectives in Production Objective: % defective to show a downward trend; in 12 months aim to reduce defectives by 6%		
QC Manager	**Production Supervisor**	**Production Operator**
• Within 4 weeks, assess to ensure that all operators are informed of CTQ aspects • Within 4 weeks, train operators in suitable data collection methods • Weekly, from data collected by way of QC data sheets, produce trend analysis displays and facilitate corrective action • Train operators where SPC needs to be applied	• Within 6 weeks, establish Skills Matrix for operators (to include knowledge of CTQ aspects) • Daily check that each operator: 1) Stamps the job traveller card on satisfactory start-up 2) Records in-process actions on QC data sheets 3) Stamps the job traveller card when releasing the job • Within 6 weeks, establish VM (Get help from QC dept.)	• On each start-up do checks for correct drawings, job-specific QC information, machine settings, tools, measuring instruments and first article control • Perform in-process check actions (to comprise inspection, measurements and adjustments) • Record each check action on the QC data sheet (identify problems and corrective action)

Figure 5-2: Example of deployed actions with goals and objectives

The overall goal gives (or changes) the direction to move toward, and also has an intention of changing the mind-set to adjust to and support the new direction. Goals give the purpose toward which an endeavour is directed and may not necessarily be strictly measurable.

The goal in the example in figure 5-2 is clarified in the objective: "% defective to show a downward trend; in 12 months aim to reduce defectives by 6%". When this objective was defined, five properties were considered to make it a SMART objective, i.e., to make it Specific, Measurable, Achievable, Realistic and Time-based.

As can be seen in the example, for different functions in the organisation and at different levels, there are actions having goals and objectives that support the overall goal and objective; and all actions have results that can be monitored.

Within large industrial manufacturing organisations in China, planned strategic requirements were in evidence in the objectives set at top management levels. These appeared to focus on key factors such as total profit, proceeds of sales, cost/sale ratio, technology investment ratio, and rarely some specific aspect of quality. Aspects of customer satisfaction, for example, on-time despatch to customers and reduction of customer complaints, which would be of strategic importance to the better Western companies, were not in evidence.

Planned objectives that were set at other management levels were not, strictly speaking, objectives; they were a mixture of year-specific work tasks and general work tasks. Examples of these are as follows:

- Ensure that production is balanced and completed per plan
- Enhance cash flow
- Ensure on-time supply from suppliers
- Monitor the Testing Laboratory quality accreditation system
- Urge other managers to complete the job on time
- Ensure the effectiveness of the quality management system
- Continue to conduct the quality improvement programme
- Control the quality of subcontracted material
- Investigate the typical quality problems
- Co-operate with design, process, and production departments, identify the gap, and try to solve the key problems

In Western organisations most of the above would be regarded as routine job duties and, as such, would be specified in job description documents.

Key Performance Indicators (KPIs):

It is commonplace in Western organisations to find important business performance goals or objectives assigned Key Performance Indicators (KPIs). These are linked to and directly support critical success factors in strategic goals, objectives and performance improvement initiatives which are different from one company to another.

Table 5-2 contains examples of KPIs which includes quality management related KPIs associated with typical business performance subjects such as profit and customer satisfaction.

Usually between five and eight KPIs are agreed with each of the management and senior personnel. Incentive is provided to motivate them to take action, and they are held accountable for the achievement of each Key Performance Indicator.

Through the regular review and monitoring of KPIs, progress is shown toward realising the company's objectives or strategic action plans, which, if not properly performed, would likely cause losses, or jeopardize achievement of the organisation's vision.

Table 5-2: Examples of Key Performance Indicators

Typical business performance subjects	Key Performance Indicators
Profit Before Interest and Tax (PBIT)	• Improvement of planned PBIT • Reduction of costs wrt. sales (cost/sales ratio) • Reduction of Price of Non-Conformance (PONC)* • Improvement of cash collection efficiency (Reduction in number of debtor days)
New sales orders	• Number of new orders versus plan
Capital employed	• Reduction in total working capital • Reduction of company inventory

Customer satisfaction	• Time to respond to order enquiries • On-time despatch to customers • Reduction of technical and commercial complaints • Reduction in turnaround time for repairs • Achievement of stated product reliability • Reduction in warranty claims • Response time to customer complaints • Number of customer satisfaction reviews performed
Health and Safety	• Safety risk assessments performed • Accidents per 10000 hours
Design and Development	• Design reviews completed and risks identified and removed • Reduction in number of engineering changes • Number of weeks project slip
Production performance	• Improvement of key plant OEE** • Reduction of repeat problems in production • Performance against scheduled delivery date • Reduction of number of defects • Reduction of concessions raised
Quality management system effectiveness	• Internal quality audits performed • Response and effectiveness closing identified corrective, preventive and improvement actions
Employee skills and competencies	• Skills and competencies reviews performed • Training provided per skill gap identified
Suppliers/sub-contractors	• Reduction in number of suppliers • Sub-contractor assessments performed • Improvements in supplier performance
*Chapter 8.3 – "Price of Non-Conformance" reduction programme **Chapter 8.2 – Enabling, supporting and improving operational performance	

In an industrial manufacturing company in China, pioneering work was found in the establishment of Key Performance Indicators; however, it seemed that the logic in determining the KPIs was quite different to that of Western organisations:

- The performance objectives were more work duty related than critical factor related and therefore the linking to high level strategic performance objectives was somewhat vague.
- The performance objectives had measurements that were generally subjective or a matter of opinion rather than having measurements that were objectively relevant, quantifiable, specific, actionable, timely and attainable.
- Many department heads passed their KPIs directly down to the next level rather than discussing and agreeing appropriately derived performance targets with the subordinate. It was also found that some lower level individuals were tasked by HR to write their own KPIs. These approaches defeat the purpose of KPIs.

The development of KPIs in these companies appears to have been influenced by their long institutionalized "Work Objective Deployment" process; this set out the yearly work objectives for the various hierarchical levels of the organisation. It seems that the yearly work objectives were developed initially from a bottom-up approach and then adjusted, especially at the higher hierarchical levels after their review by top management, to incorporate strategic requirements such as total profit, proceeds of sales, cost/sale ratio, and technology investment ratio. The strategic requirements were more in evidence in the higher level Work Objective Deployment forms, and only at this level did strategic requirements have clear measureable objectives.

In Western companies the Key Performance Indicators are part of a Personal Performance and Development Review System; this system typically operates in the following manner:
- The planning and determination of the KPIs for the subordinate is done by the subordinate's manager setting performance targets for the subordinate. These are discussed with the subordinate to ensure his complete understanding.
- The subordinate applies himself to achieve his performance targets.
- At a predefined time-period, as determined by the nature of the job and its impact on the organisation's critical success factors, the

subordinate's performance against his KPIs is reviewed. This review is performed by the subordinate's manager by way of a formal "Personal Performance and Development Review".

- The review of the subordinate's achievements with respect to his performance targets is used as a motivational means as well as to draw out actions to improve or develop his performance; it is also used to identify and remove obstacles standing in the way of his performance – e.g., lack of knowledge or resource issues. The review could also lead to a redefining of a key performance indicator.

The KPIs in the companies in China extended to fairly junior levels whereas, in Western organisations, the KPI and Personal Performance and Development Review System are not customarily applied to junior levels. Instead, organisations may employ a skill and task driven "Performance and Competency Review" System. The focus, for instance, of a "Performance and Competency Review" for a production operator, would be on the operator's skills, abilities and competency to perform assigned tasks and would include quality related skills and abilities, i.e.:

- The ability to check start-up requirements and controls, e.g., correct drawings, tools, and machine settings, availability of measuring instruments, and the correctness of materials provided.
- The skill and ability to perform process measurements.
- The skill to detect problems and the ability to respond appropriately to correcting problems.

With this approach, the operator is better equipped to prevent quality problems and can then support a higher level objective given to the workshop such as "Reduce repeat problems in production".

5.4 Internal communications for effective quality management

Internal communication is the lifeblood of any organisation; through internal communication, employees get most of the information that they need to perform their jobs, and this is obviously needed in order to establish an effective quality management system. Therefore, ISO 9001 (GB/T 19001) holds that it is the responsibility of top management to ensure that there are clear and effective communication processes established that result in people having information to do their work, and that these processes ensure effective communication of quality requirements, objectives and achievements.

There is both formal and informal internal communication. Meeting minutes, reports, notices, policies, job descriptions, organisation charts, procedures, electronic workflows, and process instructions are examples of formal communication. The exchange of ideas in the corridor and the friendly chats that develop personal relationships, are examples of informal communication.

Measurement is also a form of communication. For example, the work performer recording the number and types of defects occurring during his work process is communicating information on the issues that need action and which issues need priority attention.

In table 5-3, some important reasons for internal communications are provided, and for each reason, examples are given of the typical types of internal communication that supports effective quality management.

Table 5-3: Internal communication that supports quality management

Reason for internal communications	Examples of communication that supports effective quality management
To provide employees with the information they need to do their jobs effectively	Drawings; procedures; process and work instructions; quality checklists
To provide employees with clear standards and expectations for their work	Technical drawing specification tolerances; process control limits; performance targets; quality standards (visual and non-visual)
To make sure employees are informed about anything that concerns them	Management briefs and information notices concerning quality achievement as well as failure; electronic workflows for, e.g., customer complaint handling; quality performance measurement information
To give employees feedback on their own performance	Personal Performance and Development Review; Performance and Competency Review

To provide employees emotional support for difficult work	Supportive communications that ranges from giving encouragement to counselling (To alleviate work related stress and frustrations, and address consequent problems causing quality errors arising from distraction and lack of focus)
To allow employees to understand the situation in their work area and in the organisation	Pareto Charts, Histograms, and Defect Concentration Diagrams showing occurrence and prominence of defects; quality key point charts
To help employees maintain a shared vision and a sense of ownership in the organisation	Discussion, notices etc., to remind employees of the company vision and quality goals; sharing with employees customer feedback on quality; giving quality achievement excellence awards

In Western organisations certain basic ideas are generally taken into account by top management (and employees) when communicating:

- Communication is not one-sided: The idea is to ensure that information is understood in the way it is meant to be understood.
- Direct communication is preferable: Indirect communication may sometimes be necessary but it always carries the possibility that its message will be distorted as it goes through the network between the source and the recipient.
- Communication style must be appropriate: Cultural background, education, and perception of who has power all influence how information is sent, received and interpreted.
- Communication is through actions: Top management is closely watched by employees so everything they do is important.

Meetings can be participatory or message delivery. There appeared to be a preference among senior managers in organisations in China to convene message delivery meetings in which comment was not invited. Rarely experienced were meetings where attendees were required to respond to action items; often the concept of accountability of an individual for a particular action was unclear.

In a number of organisations in China, standard operating procedures were found that had been developed by engineers. These were also called process procedures and were similar to those seen in countries such as England and Germany. They were excellent in clearly describing and illustrating process settings, conditions and actions to operators, and were explained to be used for training purposes. This is a very good form of communication. They were locked away in cabinets after training and it is considered that better use could have been made by making these documents available in the production environment at process information stations or perhaps developing key point charts from them to be displayed at operators' workstations.

Experience was that access to a lot of operational information in industrial manufacturing organisations in China was often fraught with difficulties. Information tended to be held on personal computers, or files exclusively controlled by individuals. When the individual was not in office, e.g., away on a business trip, the information could not be accessed until his return. The individual may have been away for several days. This is quite an unthinkable situation in Western organisations as inaccessibility of company information to those who need it detracts from business transactions and performance. Typically, organisations operate central file servers, and access, storage and editing permission are electronically pre-set.

An improvement in internal communication in some manufacturing organisations in China was found by way of the implementation of "Office Automation" (OA) systems. (The backbone of an OA system is a company managed local area network which allows users to transfer data and mail across the company network, and to electronically store data. Some organisations configure the OA for action authorization by way of simple electronic workflows.)

A few organisations in China had an intranet to communicate information such like company news, company notices, and events; some organisations even communicated inspection reports.

In addition to the previously mentioned central file servers designated for data communication, the following communication arrangements are found in the better Western manufacturing organisations:

- Electronic workflow practices extend to document version control to ensure that team members working on different aspects of a design, work on the latest version of the design documents.

  The system also tracks contributions to an engineering project; this is useful to double-check changes and revert back if necessary.

  Other examples of electronic workflow are the processing of non-conformance reports for timely disposition decisions, and the efficient processing of customer complaints.

- A management briefing process where important information from top management is rapidly cascaded down the organisation by way of short "stand-up" meetings; these meetings are quickly convened by section supervisors at the various levels in the company.

  The information is read by the supervisor to his direct reports, checked to be understood, and the gathering promptly disbanded. The rule set by top management is that the information must be communicated within a very short period of time from its dispatch – as short as 2 hours.

- "Quality dashboards" are required by top management to be in place in departments and workshops which visually communicate information pertaining to product defect rate, defect cause, delivery performance, and such like. (Chapter 3.3 covers Visual Management)

- Visual aids are used to communicate acceptable and not-acceptable product attributes.

- Regular quality education is supported by top management; this is given in subjects typically embracing definition of quality, characteristics of a quality organisations, cost of non-quality, quality tools and their use, problem solving, quality problem prevention practices, quality improvement, and education in the application of processes and procedures.

- Staff and production meetings are held concerning questions such as: Are we achieving quality objectives? What repeat problems are we experiencing? How can we remove problems from a certain process?

- Employee personal performance and development reviews, and operator skills and competency reviews, are performed as mentioned in Chapter 5.3 under Key Performance Indicators.
- Quality improvement suggestion boxes are used by some companies.

5.5 Use of information to facilitate quality improvement

In the better Western companies, top management set the reporting requirement regarding quality status and quality improvement actions. This keeps them informed and also facilitates ongoing quality improvement. The reporting requirement is specified and applies to most areas of the company. It promotes the disciplined collection and analyses of useful data as well as the acting on information to bring about quality improvement.

Company-wide quality information collection and analysis systems in organisations in China, as needed for overall quality improvement, were generally quite limited. Mostly, information collected concerned scrap, quality rejects, concessions granted, and in-service product failure. Reject notices contained information on product non-conformities found during final inspections, and fairly detailed information was held in files on customer complaints arising from in-service failures. This is a good beginning but represents only a part of the information that should be collected, analysed and used for overall quality improvement:

- Firstly, the focus only on product non-conformity detected at final inspection, and scrap and in-service product failure, limits quality improvement endeavours. Information that is required for quality improvement must also come from products that are in-process, and from the various company processes – not only production processes – and must be collected from many areas of the company, i.e., from sales and marketing, customer service, design, production, quality control, logistics, installation, accounts, etc.
- Secondly, the use of information such as that indicating design weaknesses or product non-conformances, or untimely response,

must not be delayed; timely feedback to a specific individual would enable the individual to take action to correct the problem.

It was observed that quality-related measurement and information collection systems that had been initiated in organisations in China were not readily accepted and subscribed to by all managers. There was also the tendency to believe that once qualified, processes were "perfect". Scarce attention was given at the workstation to record and report naturally occurring operating, handling, quality, and tooling problems, and operators not producing the specified quota were penalized. When problems occurred, employees did not have the time to understand and eliminate the causes. To avoid lengthy stoppages, they developed quick fixes and walk-arounds. These often caused more issues, sometimes extra-operations and planned increases in inventory.

A few managers could see the benefit of collecting and analysing information for quality improvement purposes but there often appeared to be an issue with the complete integrity of this information. Employees appeared to resist accurately recording and reporting bad quality information possibly due to a blame and penalty culture, and suspicion that the information could be used against them.

This situation is different where employees find a sense of purpose and also accomplishment in measurements and the use of information. This follows management working with employees to develop measurements that have meaning for both employee and the company. Quality tools play a most necessary role in the meaning that can be extracted from collected information, and therefore education of employees in the use of Basic Quality Tools is very effective in making advancements in quality improvement. (Chapter 11 covers quality tools.)

An engaged employee, in whom a positive attitude has been nurtured towards the organisation and its values and who is aware of the business context of quality information, works with colleagues to improve performance within the job for the benefit of the organisation.

It is found that when information within the system captures management's attention, and management see its usefulness and benefits, their support and demand for its collection and routine use will usually follow. This can be the catalyst agent for the quality management system to transcend into effectiveness and for the management of quality to improve.

The use of quality information is further covered in Chapter 7.4, and information that must be targeted to reduce the Price of Non-Conformance, is covered in Chapter 8.3.

Questions for Chapter 5

5-1: Briefly describe the emphasis in the responsibilities of top management made clear in the latest revisions of ISO 9001 (GB/T 19001) quality management system requirements.

5-2: What quality management process should top management ensure is in place to make certain that the organisation has an appropriate technical solution and the ability to meet customer requirements? Briefly explain the activities of this process.

5-3: Explain how Key Performance Indicators can help to achieve the strategic goals, objectives, and performance improvement initiatives of an organisation, including those that are quality management related.

5-4: Identify quality-related performance indicators specific to each of the following seven concerns of a business: customer satisfaction, profitability, product conformance, product design and development, equipment, personnel, quality management.

5-5: Provide at least five reasons why internal communication is important to a company, and substantiate each reason with examples of typical types of internal communication that supports quality management.

5-6: (a) What areas in an organisation can provide information that can be used to facilitate quality improvement?
(b) How can management encourage collection and use of information to make advancements in quality improvement?

PART 3: COST & RISK

Subjects, practices, concepts and techniques contained in Part 3

Chapter 6: Quality costs and associated activities
- ◆ Cost of controlling quality = Prevention + Appraisal Costs
- ◆ Cost of failure to control quality = Internal + External Failure Costs
- ◆ Prevention and Appraisal activities for the control of quality
- ◆ Failure activities occurring as a result of failure to control quality

---

Chapter 7: Practices having major bearing on cost and risk
- ◆ Quality planning
- ◆ Fire-fighting
- ◆ Design review
- ◆ Design analysis and design risk assessment
- ◆ Inspection and test
- ◆ Collection and use of quality information
- ◆ Quality training and education
- ◆ Internal quality auditing

6. QUALITY COSTS & ASSOCIATED ACTIVITIES

An appreciable part of the costs that a business incurs in bringing a product or service to market is the costs of ensuring that the customer is provided with an acceptable quality. These are termed Quality Costs. They are comparable in importance to other costs in an organisation such as labour costs and material costs and have a significant impact on profitability and competitive position.

Quality Costs are incurred in pursuance of "quality of design" and "quality of conformance", and when failures occur. Because poor "quality of conformance" *can* exist, Prevention Costs and Appraisal Costs are necessary. Prevention Costs involve the costs of ensuring that design risks have been considered and eliminated, that the design meets customer and regulatory requirements, and that the production process has been planned and organised to ensure product conformance. Appraisal Costs are the cost of testing and inspecting the materials and the product.

The other types of Quality Costs deal with the costs incurred because poor "quality of conformance" *does* exist; these are Internal Failure Costs and External Failure Costs. Internal Failure Costs involve such things as the cost of rework of defective items and the cost of downtime due to failed products and materials. External Failure Costs involve the costs when failure occurs in the hands of the customer, the cost of warranty service and replacement, and the cost of product liability.

The Internal plus External Failure Costs are often termed the "Price of Non-Conformance" (PONC). The difference between the price paid and the cost incurred is the profit the business makes when the item sells, PONC is paid by the business and so comes directly off the profit that could be made by the business.

The costs of failure to control quality could arise anywhere in an organisation. For instance, there may be product design issues that begin in the design and development department, manufacturing problems could create product defects, the procurement department may acquire sub-standard components that result in product flaws, and the customer order may have been incorrectly entered so that the customer receives the incorrect product, or the quantity is incorrect, or the product is shipped to the wrong address. These issues all result in failure costs and therefore attention needs to be directed at controlling quality. Discussed in further detail in this chapter are:

1. Activities employed to control quality
2. Activities occurring as a result of the failure to control quality

6.1 Activities employed to control quality

Prevention and Appraisal activities make up the costs to control quality.

Prevention activities are aimed at preventing, avoiding or minimizing poor quality in products and services.

Table 6-1 contains some examples of Prevention activities and practices.

Table 6-1: Examples of Prevention activities and practices

- quality management system development and its continual improvement
- new product design reviews
- creation of quality plans
- development of key point inspection charts and visual aids
- improvement of manufacturing processes
- supplier and sub-contractor quality assessments
- technical support provided to suppliers
- process capability studies
- investment in quality-related information systems
- internal quality auditing (to prevent potential quality problems from occurring)
- data gathering, analysis, reporting and use
- quality improvement projects
- Statistical Process Control
- quality education and training

It is common for the better Western enterprises to "invest" in prevention by establishing activities as described in table 6-1, and by setting up a quality engineering function whose job is quality prevention-focused. (The work of the quality engineer is outlined in Chapter 4.3.)

Findings and observations indicated that little seemed to be done in manufacturing organisations in China with an appreciation that PONC reduction benefits could be derived from "investing" in prevention activities; of the prevention activity and practice examples given in table 6-1, only about 20% of these were seen to be regularly practised.

There is great opportunity for significant PONC reduction in manufacturing organisations that invest in prevention activities; for further details see Chapter 8.3 "Price of Non-Conformance reduction programme".

Appraisal activities are employed in identifying defective products before they are shipped to customers. Appraisal activities are associated with inspecting, measuring, evaluating and auditing products or services to assure conformance to quality standards and performance requirements. Table 6-2 contains examples of Appraisal activities and practices.

Table 6-2: Examples of Appraisal activities and practices

- inspection and test of purchased material, at goods receiving and at source
- laboratory inspection testing
- production process set-up checking
- inspection during first-off and in-process
- process control measurements (Statistical Process Control)
- final product inspection and testing
- field testing and appraisal at the customer's site
- supplies and product used in testing and inspection
- internal quality auditing (to verify quality management system effectiveness)
- calibration of measuring and test equipment
- outside certifications

In manufacturing organisations in China, the most commonly observed arrangements for appraisal activities with regards to the serial production of machined parts were (a) "first-off" inspections were performed during the set-up stage at the start of a production, (b) during production very few inspections were witnessed as being performed and no in-process control measurements were recorded, (c) sometime after production had been completed, final inspections were performed. This placed a lot of emphasis on final inspections and tests to screen out defective products and parts.

Final inspections were hardly ever seen to be performed by operators; typically, dedicated inspectors were employed to perform final inspections either on a sample or 100% basis. This approach does not support dynamic quality control as discussed in Chapter 4.1.

In Western organisations, machine operators perform first-off and in-process inspections and measurements. Work is released by them from their work-area only when they are satisfied that it meets quality requirements. Machine operators are responsible for acting quickly during production to correct out-of-limit measurements, excessive variation, and unwanted machining related issues such as burrs and score marks. A limited number of patrol inspectors are frequently employed; their focus is more on prevention actions to "start right and stay right".

Regarding incoming goods in industrial manufacturing organisations in China, customarily receiving inspection was routinely performed. Some organisations would send their inspectors to supplier's premises to conduct inspection before the product was shipped to the factory; this was especially the case when large and/or complex product was sourced.

In Western organisations, receiving inspection may or may not be performed. This could depend upon quality assurance activities at the supplier's premises which may have included in-process monitoring and product release at their premises by the purchaser's quality assurance representative, or that the product has the qualification status of "Ship-to-Stock". Supplier Quality Assurance is discussed in Chapter 9.

6.2 Activities occurring as a result of failure to control quality

Activities occurring as a result of failure to control quality are Internal Failure and External Failure activities.

Internal Failure activities result when products or services do not conform to requirements or customer needs prior to delivery to the customer, and from unsatisfactorily completed business tasks that are required for delivering customer satisfaction.

Table 6-3 contains some examples of Internal Failure activities and practices.

Table 6-3: Examples of Internal Failure activities and practices

- net scrap
- rework labour and overheads
- re-inspection of re-worked product
- re-testing of reworked product
- extra operations (touch-up, de-burring, re-cutting of assembly holes, etc.)
- downgrading
- failure analysis activities to establish product failure cause prior to delivery
- expediting
- downtime caused by quality problems
- disposal of defective products

In the better manufacturing organisations in China it was found that the Internal Failure Costs of scrap and rework costs were recorded and reported. This in reality represents a small percentage of PONC. Activities such as de-burring, re-cutting assembly holes, product downgrading (reducing the selling price due to poor quality), and expediting were not acknowledged as Internal Failure Costs; it was quite normal to find workstations dedicated to such operations as de-burring, and many staff chasing late deliveries and poorly conforming products from suppliers.

The latter activity attracted what is referred to as "expediting costs". Expenses incurred in expediting can be by way of the employment costs of people brought into the organisation to expedite, and the costs of overtime and/or express transport. Practically all expediting costs are as a result of failure to control quality and arise from activities such like;

- chasing unreliable suppliers to ensure that goods, services or sub-contracted work meet specifications and arrival times;
- attempting to recover time due to production delays arising from rework;
- attempting to meet delivery dates that are in reality unrealistic but have been agreed to obtain an order.

In Western organisations such activities are targeted for reduction and elimination; this is discussed in Chapter 8.3

External Failure activities result when products or services do not conform to requirements or customer needs after delivery to the customer.

Table 6-4 contains some examples of External Failure Cost activities and practices.

Table 6-4: Examples of External Failure activities and practices

- customer complaint handling (both in and out of warranty)
- complaint investigation
- repair cost of returned product or of product in the field
- customer allowances and discounts due to quality problems
- time with dissatisfied customers
- handling and investigation of rejected or recalled product, including transport
- warranty repairs and replacements
- lost sales arising from a reputation for poor product
- repairs and replacements beyond the warranty period
- liability arising from defective products

External failures can be very costly and extremely damaging to the company; the way in which these are handled, and the corrective or remedial action taken, impacts on company image as well as on immediate and long-term profitability. Urgency of response is vital; most customer complaints received by Western organisations will get prompt attention. The attention will be focussed firstly on assuring the customer that action will be taken within a planned time, and then this action will be followed through and made to happen within this planned time.

Some cases were discovered in industrial manufacturing organisations in China where the fault had not been resolved several months after it had been reported; it was ascertained that the customer-facing personnel in the Chinese organisation frequently found themselves trapped between a complaining customer and non-co-operating internal departments.

In the better Western organisations, the root-cause of the issue will be hunted down and investigated. The cause of the failure may be due to one or more of many reasons, e.g., defects in production, design shortcomings, poor handling during transportation, bad storage, or misuse or inappropriate use by the customer. Whatever these failure-causes may be, the optimum remedial action will be sought, and steps will be taken to eliminate re-occurrence.

It was observed that the manner in which external failures were handled in organisations in China could depend upon who was complaining, and to whom the customer was complaining to in the supplier's organisation. Most serious shortcomings were frequently handled by senior personnel graciously wining and dining the customer.

In the face of an important customer, irrespective of the cause of the problem, the supplier would agree to replace the product.

In the industrial manufacturing sector, both in China and the West, it is found that many products suffer abuse in the hands of customers. Western organisations will not hesitate to point this out to their customers and refuse liability. On a few occasions this type of response has been witnessed of Western suppliers towards powerful Chinese customers – the response has not been well received; the Western supplier has even been negatively labelled.

Questions for Chapter 6

6-1: What are Quality Costs? Explain the different types of Quality Costs.

6-2: Identify at least five typical activities and practices that are commonly considered to be part of normal operation costs but in fact contribute to failure costs, i.e., identify failure cost incurring activities and practices that exclude scrap, rework, customer complaints and repair.

6-3: Identify at least six prevention activities and practices that most likely will bring benefit to an organisation.

7. PRACTICES HAVING MAJOR BEARING ON COST & RISK

Risk is related to the possibility of events and activities impeding the achievement of an organisation's strategic and operational objectives. The concept of "risk" in the context of quality management and ISO 9001 (GB/T 19001) relates to the uncertainty in achieving these objectives, and in the organisation's ability to consistently serve customers with conforming goods and services.

Risk-based thinking makes preventive action part of the routine. The successful organisation strives to do as much as economically possible to prevent, avoid or minimize poor quality in products and services, and when prevention proves less than perfect, the organisation then has practices in place to identify defective products before they are shipped to customers. Prevention and Appraisal practices aimed at achieving this are outlined in Chapter 6.1.

In industrial manufacturing organisations in China, certain practices that should have significant bearing on cost and risk stand out as requiring an adjustment in emphasis to re-position their significance; these are:

1. Quality Planning
2. Design Reviews
3. Inspection and Test
4. Quality Information (Intelligence of Quality Management)
5. Competence and Quality Awareness
6. Internal Quality Auditing

7.1 Quality Planning

Quality Planning is a prevention practice; it applies to the quality management system and to the development of its processes, to process control and quality control in job planning, to continual improvement and to quality objective achievement. Quality Planning sets the stage for the quality control efforts of an organisation to avoid having defective products reach customers, and aims to eliminate the risk of poor quality and PONC.

The need for quality planning could arise from issues identified in the company's strategic planning process (discussed in Chapter 5.3), or from the management review of the organisation's quality management system. These issues may be, for instance, new quality objectives, company organisational change, or quality management system change in emphasis. The need could also arise from customer order special requirements, weaknesses found during supplier assessments, data trend analysis identifying opportunity for improvement, and internal quality audit findings – e.g., identifying risk due to major quality management shortcomings or the absence of control.

A product containing special or non-standard requirements stipulated by the customer could have these requirements identified in a Product Quality Plan or a job planning and control document such as a Process Control Job Card; this document would identify job-specific activities, verifications and controls.

Weaknesses found during an assessment of a particular supplier, or special requirements applicable to a bought-in item could give rise to a Supplier Quality Plan. (See figure 9-1 for requirements for quality plans for bought-in product.)

In industrial manufacturing organisations in China, the types of Quality Plans described thus far were not in evidence. Those seen were annually issued Quality Management Plans, but the nature of their contents was rather unexpected:

In the West, a Quality Management Plan would typically document how an organisation will plan, implement, and assess the effectiveness of its quality management system. It will describe how an organisation must structure its quality system and ensure preventive, corrective and improvement systems are in place. It will identify quality processes and procedures, areas of application, and roles, responsibilities and authorities. In Project Management, a Quality Management Plan would include processes and procedures for ensuring that quality planning, assurance and control are conducted throughout the lifecycle of the project with the purpose of ensuring the effectiveness of project work processes and the monitoring and control of project deliverables.

With this in mind, it was a surprise to find in China an annually issued document named as a Quality Management Plan that contained quality management function normal duties and routine quality operation-orientated tasks. Quality function normal duties comprised about 80% of the content of the document – these were duties such like "conduct internal quality audit", "timely issue corrective action requests", "prepare for the external quality audit" and "perform verification of key processes". In manufacturing organisations in the West, such duties would be part of the job descriptions of employees.

A further issue noted in organisations in China was that there appeared to be limited interest taken by middle management in realising actions defined in quality plans. This was also the case with the action-outputs of their organisation's quality management systems management review, and is believed to relate to a quality ownership issue.

In manufacturing organisations in the West, the ownership of the performance of a key task or a major activity is resolved through top management assigning a responsible person and allocating individuals to the implementation of the key task or major activity. The individuals involved could very likely have measures of their performance in the key task identified on their Key Performance Indicators.

Quality planning could involve traditional project planning; for instance, the following major quality objective would be arranged to be attained in a project plan, to achieve the stated quality operation objective: *"Reduce Appraisal Costs by establishing effective Operator Self-Control in the workshops, and decrease the need for the number of Quality Inspectors from 130 to 50. Time period: 24 months"*.

There would be several steps or stages in such a plan; actions would be identified that lead to stage milestones to be achieved at planned dates, and personnel responsible for the actions would be identified.

Symptomatic of a quality planning deficiency is "fire-fighting".

"Fire-fighting" involves an emergency allocation of resources required to deal with the quality problem; this is likely to tie-up resources that are needed elsewhere, and do this again and again as it recurs. Fire-fighting and last-minute reworks are difficult situations from which to exit because there always appears to be no time to stop and solve them correctly. Generally, organisations that are in fire-fighting mode indicate that they do not have a system in place that helps them improve from their mistakes.

"Fire-fighting" repetitive or recurring quality problems was disturbingly too frequently seen in manufacturing organisations in China and should have been addressed in the organisation's Quality Management Plan with activities to ensure the following:

- That preventive, corrective and improvement systems are in place and accessible at all levels in the organisation.
- That management's education in quality management (see figure 7-4) ensures appropriate leadership in, and commitment to continual improvement. This means that management are educated to take a different approach to managing quality other than penalizing employees in the face of quality problems which causes employees to not willingly disclose quality problems.
- That proper root-cause analysis techniques are known (through education and training in the use of problem-solving techniques and quality tools), and problem solving is not only handled by a select few, but by trained people throughout the organisation.

7.2 Design Reviews

The design review is a means for controlling the "quality of design" and is a risk control process. The "quality of design" is the quality specified by the designer in drawings and technical specifications, and it determines the intrinsic cost of the product.

The main purposes served by designs reviews are outlined in table 7-1.

Table 7-1: Purposes of design reviews

1. To determine if the product will actually work as desired and meet the customer's functional and performance requirements, as well as regulatory and statutory requirements, including those of safety.

2. To ensure that satisfactory reliability is achieved and improved where possible. (Endurance laboratory test results and information from reliability studies on equipment in field service are used for this purpose.)

3. To ensure that the build-up of tolerances is taken into account in assemblies and components, to obviate the need for costly customization to achieve the necessary fit.

4. To standardise on components and assemblies as much as possible to avoid the myriad of costly variations and the impact on drawing control, tooling, jigs and fixtures, measurement and test.

5. To determine if a new design is realistically economically producible, bearing in mind the resources available, e.g., the capability of the factory's equipment and the competency of available manpower.

6. To ensure that information from previous designs and feedback from actual customer use are taken into account to develop better designs.

7. To determine if the new design is maintainable and repairable, and that replacement assemblies and components are interchangeable.

8. To identify any problems arising from the aforementioned, and to propose necessary actions.

In industrial manufacturing organisations in China there appeared to be some misunderstandings of the purposes and applications of design reviews; for example, the belief that a design review is an error-check of an engineer's drawing by the engineer's peer or the engineer's chief, and the thinking that design reviews do not apply to the smaller sub-assemblies.

Design review reports were found simply stating that the design inputs have been accounted for in the design outputs. In reality, such statements have no value at all and perhaps serve no more than to indicate an attempt to satisfy GB/T 19001 quality auditors that could possibly be lacking design engineering expertise.

A discussion of observed findings in industrial organisations in China follows with the aim of highlighting areas where design reviews could be improved; reference is made to the purposes of design reviews as summarized in table 7-1:

Item 1, table 7-1, Design analysis and risk assessment:
Whereas it is common to find organisations in the West employing design analysis and risk assessment tools to identify design weaknesses, potential failure areas, and critical-to-quality characteristics, limited usage of design analysis tools and no usage of design risk assessment tools was evident. Table 7-2 contains an explanation of such tools.

Table 7-2: Explanation of design analysis and design risk assessment tools

Finite Element Analysis (FEA):

FEA is an accurate design analysis technique to predict the performance of a structure or mechanism under in-service or abuse loading conditions.

FEA is commonly used early in a design process to try out new concepts and to optimise designs before any prototypes are made and tested. FEA is often used to validate designs before committing to manufacture.

Failure Mode and Effect Analysis (FMEA):

FMEA is a common design risk assessment tool and has the objective of outlining all possible failures, their effect on the system, the likelihood of occurrence, and the probability that failure will go undetected. FMEA provides an excellent basis for classification of characteristics, i.e., for identifying critical-to-quality and other variables.

FMEA can take a "Hardware Approach" which lists individual hardware items and analyses their possible failure modes, or take a "Functional Approach" which recognises that every item is designed to perform a number of functions that can be classified as outputs. Their outputs are listed and their failure modes analysed.

Fault-Tree Analysis (FTA):

FTA is a common design risk assessment tool. It provides a graphical representation of the events that might lead to failure. In FTA, a primary event is defined (the failure condition under study), the system is examined to understand how the various elements relate to one another and to the top event. A fault tree is constructed, starting at the top event. The fault tree is analysed to identify ways of eliminating events that lead to failure.

Item 2, table 7-1, Equipment reliability:

Although information such as the duration of equipment operating time to failure and cause of failure could be extracted from individual equipment failure reports, this information was rarely found to be collected over time for examination and analysis to identify failure causes and equipment reliability improvement.

Western manufacturers are very keen to know the attributed failure causes and reliability performance of their equipment, as poor reliability would cause warranty costs and could likely effect future sales. Information from reliability performance of similar equipment in field service is used to guide the designer, and where possible, endurance testing is done on product prior to release for volume production.

Item 3, table 7-1, Build-up of tolerances:

Often encountered were difficulties caused by a build-up of tolerances. By way of an example and to serve to explain other issues arising from inadequate design review, a particular case is cited:

A team of operators were observed enlarging mounting holes in crusher frames with a flame-cutter. This was necessary due to the accumulated tolerances of stack-height adjusting plates between the crusher pan and crusher frame. For each assembly, this work took three men two full days. This had been standard work for seven years.

The work was identified by a fresh pair of (Western) eyes to have arisen due to poor design engineering and an inadequate design review process. Had proper tolerance build-up been conducted by the designers, this extra-operation would have been avoided. Not only did the extra-operation waste time and money, but the flame-cutting process left an ugly aesthetic appearance.

An investigation was launched and changes were made. The design office changed the mounting holes in the crusher frames into slots which removed the need for this extra-operation.

Item 4, table 7-1, Standardisation of components and assemblies:

The design problem described in 3 above was found in 26 types of crusher frames. The change involved 416 drawings. The large number of drawings was the result of no standardisation in configuration being practised which resulted in many similar drawings being generated for the same assembly over the seven years.

Item 7, table 7-1, Interchangeability:
The design correction to the problem described in *Item 3* above solved a long-standing and costly interchangeability issue that occurred when the crusher pan had to be replaced or changed. In addition to eliminating the interchangeability issue and saving the extra-operation cost involving 48 man-hours per assembly, the design correction reduced time-to-deliver, and also removed the "eyesore" that flame-cutting had produced.

Item 5, table 7-1, Realistic and economically producible design:
There have been many occasions in Chinese organisations when the engineering tolerance specified for components was found to be unrealistically and unnecessarily tight. This sometimes resulted in the production machines having difficulty maintaining these tight tolerances. Another example is that of in-house standards for chemical content of cast material copied from national standards, but tolerances had been tightened in the in-house standard without reason that could benefit the cast material's properties. These tightened tolerances attracted many unnecessary reject notes which regularly forced delays and the accumulation of work-in-progress due to the lengthy disposition decision processing time.

(The reason given for the tightened in-house standard was the belief that this would improve the discipline of control in the foundry shop.)

Item 6, table 7-1, Design improvements:
Despite the wealth of data captured in reports arising from in-service product failure investigations, it was found that this data was rarely analysed to extract useful information that could be used to improve existing designs or to develop better designs.

The following generally holds true as good practice guidelines for the application of design reviews:

- Designs subject to review include parts, sub-assemblies and main assemblies.

- For simple designs it may be necessary to have only one formal design review at the very end of the design process; however, performing only one formal design review is considered to be very risky for more complex designs. Not only is there a probability that important considerations could be overlooked, but, if there are any problems identified as a result of the design review, it may be very costly and in some cases too late to go back and redo some design activities in order to correct problems.

- Design reviews are normally conducted at various points in the design process. Typically, design reviews takes place,
  - upon requirements definition, conceptual design or on preliminary design sketches,
  - on detailed drawings, or after prototypes are designed,
  - after prototypes are built and tested, and,
  - prior to release of design into volume production

- Depending on who participates in a design review, it can be classified as formal or informal.

  An informal design review involves those individuals directly involved in the design.

  Formal design reviews include applicable subject matter experts who are not directly involved in the design but are competent to review and comment on the design.

  Formal design reviews are effective in preventing problems in a later stage particularly when all relevant parties are involved; this includes internal departments and functions as may be appropriate, as well as customers and subcontractors.

  Some of the types of formal design reviews are described in table 7-3. Each formal design review is regarded as a baseline in the design process from which to progress, and each of these baselines has criteria against which the design is assessed.

Table 7-3: Types of design review

Preliminary Design Review:

This review often includes or is preceded by a Requirements Review to ensure that all of the appropriate requirements and design constraints have been clearly identified and allocated to the assemblies. In the case of large systems being developed, the allocation of requirements to individual items or assemblies could be examined in a System Design Review.

In a Preliminary Design Review, the relative merits of different concepts are evaluated for technical adequacy and general compliance with requirements. Assumptions and calculations are provided, and whenever possible, preliminary design sketches or preliminary prototypes are used to communicate the various concepts.

The budget, schedule, and potential risk items are highlighted.

Critical Design Review:

A Critical Design Review is conducted to evaluate the design against the detailed requirements, and occurs after the detail design is complete and prior to the fabrication of prototypes or pre-production models.

A Critical Design Review includes a review of calculations used in the design, progress of the project, and management of risk. It also includes examination of the test plans that will be applied to prototypes or pre-production units to verify the design against the requirements. A Production Readiness Review is often included (see below).

Final Design Review:

A Final Design Review is conducted after prototypes or preproduction units have completed verification testing. Problems encountered during this testing are examined, and respective solutions are formulated.

Necessary changes to the product with respect to performance, cost, reliability and manufacturing issues are agreed upon prior to the initiation of volume production.

Production Readiness Review:

Production Readiness Reviews are particularly critical for products of which a large number will be produced. These reviews may be held throughout the development of a product. In the early stages of a design project the review could concern, for instance, component sourcing and high level manufacturing issues, and as the design matures the review would become more concerned with the details of manufacturing.

Before permission to initiate volume production is given, all questions on a planned production readiness checklist must be answered satisfactorily.

Design review baseline checklists are found in use in leading Western companies; a "Production Readiness Review" baseline checklist would contain questions, for example, as follows:

- Have the critical-to-quality characteristics been identified, e.g., concerning material properties, size dimensions, surface finishes?
  Note: Identification of critical-to-quality characteristics cannot be done (a) after release of drawings to Production or Sub-Contractors, and, (b) cannot be assumed to be performed by an Inspection Department who do not have insight into the design criticality or design constraints.
- Has an FMEA or an FTA been conducted? See table 7-2.
- Are test results available and satisfactory?
- Have manufacturing concerns been addressed such as the availability of suitable production machines, the ease of manufacture, the handling of the product components during production?
- Have drawings been placed under document control and have they been checked and approved?
- Have parts lists been specified, placed under document control, checked and approved?

Effective design reviews play a huge role in controlling risk, and in controlling the "quality of design" which defines the intrinsic cost of the product, and in foreseeing potential quality problems and preventing them.

7.3 Inspection and Test

In an attempt to screen-out quality problems, end-of-process inspection and testing was frequently seen to be practiced in industrial manufacturing organisations in China; for instance:

- In a machine workshop at the production set-up stage, measurements were performed, but once set-up was deemed to be satisfactory, measurements were then typically performed at the very end of the production process by an independent inspection function.

- In a heat treatment workshop, inspectors conducted random surface hardness checks after the oven batches had cooled down.
- In a welding workshop, Non-Destructive Tests (NDT) were carried out on some randomly selected articles.

There are problems with this way of working that could have a very bad effect on cost and risk. Not only does the significant amount of effort given to end-of-process inspection and test lead to high Appraisal Costs, it also fails to prevent internal failures therefore Internal Failure Costs are high, furthermore, end-of-process inspection and test fails to screen out all internal failures therefore the risk of external failures will increase, so the overall cost of quality is high.

The issue is that inspection *in* and *of* the process is neglected so there is practically no focus on process control.

Considering the machining of parts, i.e., drilling, boring, milling, cutting, broaching – at any time of the machining process, faults of a "special" or "common" cause could occur. "Special" causes could be due to, e.g., random machining debris slightly altering the work piece set position, the machine needing adjustment to maintain a specified tolerance, random imperfections in material, or the operator not paying attention. "Common" causes could be due to, e.g., inappropriate procedures, excessive machine wear and tear, or variation of material properties. (See Chapters 10.1 and 10.2 for further information.)

The lack of in-process inspections during the machining process means that all but the obvious defects could go undetected. 100% inspection at the very end of production has a probability of not detecting all defects; this is because defective items will slip through unless the inspection process is itself 100% effective, and this will, by the nature of human and measurement errors, have a probability of uncertainty. It also is often difficult to pinpoint the specific cause of many defects when parts are found defective at the very end of production by an Inspector and therefore difficult to ascertain accurate corrective action, so the production of defective items continues.

Heat treatment and welding are "special processes", as are the application of surface finishes such as powder epoxy painting, hot-dip galvanising, and plating. "Special processes" refer to processes where compliance with requirements cannot be determined by inspection of the finished article, and also where a part of the material property cannot be assured by inspection during processing. Items emerging from a "special process" commonly rely on Non-Destructive Tests (NDT) to verify their acceptability, but NDT has limitations, for example:

- The often encountered measurement of surface hardness limits the verification of heat treatment effectiveness to the surface of the article. This would be unsatisfactory for articles requiring uniform through-hardness. It should also be borne in mind that the measurement is only relevant to the article that was situated in a certain location within the batch of articles that received heat treatment in the oven.

- NDT cannot detect all welding associated defects, for instance, it cannot detect problems occurring in the heat affected zone (HAZ). Only once the fabricated item is put in use will stress related cracks become evident adjacent to the weld seam.

- Some characteristics of a painted surface, i.e., colour, gloss, dry film thickness, and the absence of visual paint defects, can be immediately detected at the end of the paint process. However, paint defects related to poor paint adhesion and corrosion resistance become apparent only after the product is in use, only then will it become known if the painted surface will chip or rust. The only options available immediately after "special process" treatment to determine the integrity of paint surface adhesion and corrosion resistance qualities require destructive testing.

Attention therefore needs to be given to the control of the "special process", and the equipment used, and the proficiency of personnel attending to the operation of the "special process".

A further problem in not taking steps to prevent problems during the machining, heat treatment, welding, painting, etc., is that, once the products are processed, they have "added value", and their fate, if found defective, will need to be carefully considered.

This often results in a protracted disposition cycle to determine what to do with the non-conforming work, and the event diverts personnel from productive work and results in a build-up of work-in-progress.

Self-inspection and "Operator Self-Control" typically form the first line of defence in stopping problems as close to their source as possible. In the better organisations, this quality self-control is encouraged to be applied in every process, from order intake, through production, to dispatch and customer invoicing; the intent is to detect the problem as early as possible to avoid subsequent wasted "value-added" costs.

When, Taiichi Ohno[9] undertook a study tour of Ford in Detroit at the beginning of the 1950s, he came to the conclusion that there was too much waste and rework in the then so-called most efficient car plant in the world. His arguments were as follows:

"Quality comes not from inspection, but from control and improvement of the process. Moreover, mass inspection has a demoralizing effect on employees, which reduces the likelihood of zero-defect production. The mass-production practice of passing on errors to keep the line running caused errors to multiply endlessly. Every worker could reasonably think that errors would be caught at the end of the line and that he was likely to be disciplined for any actions that caused the line to stop. The initial error, whether a bad part or a good part improperly installed, was quickly compounded by assembly workers further down the line. Once a defective part had become embedded in a complex vehicle, an enormous amount of rectification work might be needed to fix it. And because the problem would not be discovered until the very end of the line, a large number of similarly defective vehicles would have been built before the problem was found."

---

[9] Taiichi Ohno (February 29, 1912 – May 28, 1990), Japanese businessman who followed the principles of Deming and Juran to become the architect of Toyota's Production System (TPS).
Lean Manufacturing's origins can be traced to TPS.

This could have current relevance to the many manufacturing plants in China that rely heavily on end-of-process or final inspection.

W. Edwards Deming[10] stressed that relying 100% on final inspection of machined parts, sub-assemblies and components, is equivalent to acknowledging that the product creation process does not have the capability required for the product specifications.

Certainly in the better industrial manufacturing organisations there is more emphasis on controlling quality during the product creation process where the means is provided for work performers to perform quality self-control during product creation or build.

For quality self-control to be effective, training and the means to make self-control possible must be given. On-the-job skill training and quality education must be provided. The means given would depend on the specific job but could include what and when to check on verification check lists or key point charts, and how to check may be provided by way of instruments to perform the inspection or test.

A standard practice employed in the better organisations is for the internal customer to verify the inspection status of the article that is passed to him for further processing. The internal customer is not expected to thoroughly inspect the articles received but rather to examine that proof of inspection status is provided by his internal supplier in some distinct way, e.g., on the job traveller or on the article.

7.4 Quality Information (Intelligence of Quality Management)

Quality management cannot reach any level of effectiveness without the continuous use of quality information; indeed, quality information could be considered as the intelligence of quality management. This intelligence is used to ensure customer satisfaction and quality

---

[10] W. Edwards Deming (October 14, 1900 – December 20, 1993), American engineer, statistician, professor, author, lecturer, and quality management consultant.

management system effectiveness; it is used to guide and improve human performance, product design, product conformance and customer service, and it is used to identify risk and where efforts need to be placed to reduce PONC.

Data collected must, after timely analysis, provide a factual basis for making decisions through a focus on objective information about what is happening in the process, rather than subjective opinions. Objective information can come from analysis of, for example;

- in-process production measurements;
- response times, e.g., in closing customer complaints;
- investigating reasons why on-time delivery is not achieved;
- operation of in-service equipment (to determine product reliability, MTBF);
- regular internal quality audits

Key Performance Indicators indicate what needs to be measured to reach quality objectives and goals. Table 5-2 in Chapter 5.3 contains examples of Key Performance Indicators from which it is clear that information must be collected from many areas of the organisation, e.g., from sales and marketing, customer service, design, production, logistics, field service, accounts, quality control and human resource.

It was found that quality data was collected in industrial manufacturing organisations in China primarily from end-of-process inspections and tests. There was no measuring of, for instance, response times in handling customer queries, on-time delivery achievement, in-service equipment performance, responses to actions arising from internal quality audits, or response times in handling customer complaints.

Generally, the end purposes for the collection of quality data seemed more focussed on working out loss through scrap and the calculation of employee monetary penalties.

In the better organisations that recognise the power of the effective use of quality information to control quality, solve and prevent problems, and to effect quality improvement, the following principles are found that guide the targeting, collection and use of quality information:

- Measurements and standards are set on customer wants and needs including response time to queries, delivery, and service, and achievement of reliability performance.
- Quality measures are aligned with business objectives.
- Networks for communicating quality information are established and maintained.
- Awareness of current quality status is made clear in standardised quality reports and on quality information dashboards located in prominent places throughout the organisation.
- Use of quality information focuses on correcting the process that contributed to failure rather than installing short-term fixes to problems.
- Corrective action plans are based on root cause analysis.

The reality is that most quality problem solving and quality improvement often requires a lot of work; this essentially involves:

- deciding on what data would be relevant for use as quality information,
- identifying collection points for this data,
- designing data collection methods,
- arranging for the data to be captured and communicated for use and analysis, then,
- analysing the data to reach intelligent fact-based decisions

This is indeed time-consuming work and requires the help of specially trained personnel with knowledge of quality tools; these personnel were generally not available in industrial manufacturing organisations in China.

7.5 Competence and Quality Awareness

The competence and quality awareness of persons doing work under the control of the organisation have unquestionably a significant bearing on the organisation's performance. The 2015 version of ISO 9001 (GB/T 19001), the International Standard for Quality Management System Requirements, stipulates the following requirements specifically for competence and awareness:

Competence requirements (clause 7.2) – the organisation must:
a) determine the necessary competence of persons doing work under its control that affects its quality performance;
b) ensure that these persons are competent on the basis of appropriate education, training, or experience;
c) where applicable, take actions to acquire the necessary competence, and evaluate the effectiveness of the actions taken;
d) retain documented information as evidence of competence

Note: Applicable actions can include, for example, training, mentoring, or reassignment of currently employed persons; or hiring or contracting of competent persons.

Awareness requirements (clause 7.3) – persons doing work under the organisation's control must be aware of:
a) the quality policy;
b) relevant quality objectives;
c) their contribution to effectiveness of the system, including benefits of improved quality performance;
d) implications of not conforming with system requirements

These requirements are indeed quite specific and apply to all persons doing work under the control of the organisation, including top management; they are intended to prevent the risk of incompetent and quality unaware employees from ruining the quality objectives of the organisation, topmost of these quality objectives is providing the customer with products or services that meets his requirements at a cost that avoids PONC caused by incompetent personnel.

Quality awareness education is an inescapable requirement for organisations wishing to attain and hold ISO 9001 (GB/T 19001) accreditation. Furthermore, and certainly for the greater majority of industrial manufacturing organisations needing people with unique technical work-specific knowledge, it is the responsibility of the organisation to take actions to ensure that people doing work under its control that affects quality performance, acquire the necessary competence.

The skills and job knowledge of employees in the better organisations in the West are complimented by quality education which aims to achieve the following:

- Impart a common language of quality and the concepts of quality to every work performer in the organisation.
- Provide an understanding to individuals of their roles in achieving quality, how their work affects quality, and the consequences of bad quality.
- Impart a comprehension of the intent of the various quality management processes and the reasons why the requirements must be fulfilled.
- Educate work performers on the need for quality improvement and how to go about achieving quality improvement.
- Provide education in the quality tools as may be needed for work performers; the education has the goal of imparting the ability to use quality tools effectively.

Quality education aims to positively influence the attitudes, knowledge, skills and potentially the behaviours of those who take part in activities. With quality educated employees, the company can be more progressive and try new things such as Statistical Process Control and Lean Methods which are geared to prevent loss.

Quality education programmes are required by top management in the better industrial manufacturing organisations in the West. Top management in these organisations include themselves in the programmes because they want to understand the language of quality and the concepts of quality; they want to ensure that everyone else is being taught the common language of quality, the concepts of quality, and what employees need to do to achieve the quality standards and objectives of the organisation. They want employees to know how quality affects the goals of customer satisfaction and the ongoing success of the organisation, and that quality-guided actions are vitally important in achieving these goals.

Given the influential position of top management in China, and the deference (meaning, respect and esteem) that employees have towards top management, it is particularly necessary that top management receive quality education in the concepts of quality, their leadership responsibility in quality management, and the strategy for quality improvement.

The quality education will help senior people understand their role in developing a prevention-orientated quality management system, how they are supposed to react to non-conformances, and how they can encourage improvement in the quality process.

The persons in the middle management group in large manufacturing organisations in China need to receive a greater depth of quality education because they have to make the quality management system and all of its processes work, and work effectively. This need is no different to similar organisations in the West. Middle managers are key persons in communicating and tracking different kinds of goals and in making information flow up and down, and need to show their quality commitment to their people every day.

Through observation and experience it is strongly believed that top management has a catalytically important role to play in the motivation and commitment of quality management to the middle management group. A tendency was noted for the middle management group to do only what was necessary to maintain the status quo and to play lip-service to quality principles. It is evidentiary to note that in in-house quality education courses given in organisations in China, if top management were present, middle management attendance was surely guaranteed; however, if top management were not present, many of the middle-managers would show their face at the beginning of the course and then quietly slip out to attend to "other business".

In the better organisations, where the values of quality management are embedded in the organisational culture, this does not happen – middle-managers share a vested interest with their colleagues in making quality management effective. Another consideration is that they are held accountable for their actions, and their Key Performance Indicators include quality goals.

Table 7-4 contains a quality education syllabus for managers. The purpose is to help to give managers a comprehension of what is necessary for effective quality management.

The intention of this education is to influence their thinking and attitude in a positive way towards a prevention-orientated quality management, and a focus on quality improvement.

Table 7-4: Quality education syllabus for managers

- Definition of quality and the characteristics of a quality organisation
- Quality management processes and practices (purpose, aims, benefits)
- The process approach
- Understanding the Cost of Quality, what are Prevention and Appraisal Costs, what contributes to Failure Costs, and the Quality Iceberg showing hidden costs
- Roles and responsibilities within the quality management system
- Requirements for effective "Operator Self-Control" (start right, stay right and end right)
- Measurement of quality and of conformance to requirements
- Handling non-conformances, root-cause analysis, and corrective actions
- Quality goals and objectives
- Promoting and supporting quality improvement, PDCA, quality improvement teams
- Introduction to quality tools (for data analysis, problem identification and quality improvement)
- The purposes of internal quality auditing and required responses to audit findings

Non-management employees receive quality education by way of various media and techniques. In the West, some of the main techniques employed are:

- Train-the-trainer, where personnel are selected and trained in subjects such as the definition of quality, measurement of quality, measurement of conformance to requirements, handling non-conformances, root-cause analysis, corrective actions and PDCA. Once qualified, these personnel can then train others in the organisation.
- Quality control teaching videos.
- Booklets containing key quality principles and practices, and examples of the application of Basic Quality Tools.
- Targeted quality education in the use of an applicable quality tool in a specific operation.
- "Operator Self-Control" actions for a specific operation.
- Visual displays for quality standards, quality targets, and quality performance measurements.
- Specialist lectures on, for example, design review practices, and configuration control.

Personnel are required with specialist quality knowledge to develop and facilitate quality practices and the improvement process. In virtually all major industrial manufacturing organisations in the West, quality engineers are employed; these personnel have an engineering education complemented with education and experience in the body of knowledge and applied technologies that includes:

- Development and operation of quality systems and quality controls.
- Methods of testing, inspection, and non-destructive testing.
- Use of statistical methods to diagnose and correct improper quality control practices.
- Understanding of human factors and motivation.
- Understanding of quality cost and risk.
- Knowledge and ability to develop and administer information systems.
- Capability to assess quality and technical processes to determine their effectiveness, to identify deficiencies and appropriate corrective correction.

Another group of personnel that requires specialist training are Internal Quality Auditors. These persons need to be trained to perform regular internal quality audits with the purpose of assessing the effectiveness of the implemented quality management system and its processes, and with the purpose of identifying risks and improvement opportunities.

Table 7-5 contains a three-module education syllabus for Internal Quality Auditors. This is the basic knowledge that the candidate would require to learn. There is also an essential skill content taught through participation in appropriate tasks and activities.

Table 7-5: Quality education syllabus for internal quality auditors

1: Introduction to ISO 9001 (GB/T 19001) Quality Management System
- Brief history of ISO 9001 (GB/T 19001)
- ISO 9001 (GB/T 19001) quality management principles
- The process approach
- PDCA
- Measures of performance of the organisation's quality management system and its processes
2: ISO 9001 (GB/T 19001) Requirements
- Clause by clause explanation of the quality requirements
3: Important understandings and aspects of quality auditing
- Definitions and meanings
- Reasons and requirements for auditing
- The quality audit process: Planning – Execution – Making an Quality Audit Report – Lead Auditor Checks – Corrective Action – Follow-up Checks
- Non-Conformities – probable causes; major/minor classification; handling of corrective and prevention action
- Pitfalls to avoid

The scope and depth of quality education outlined in this chapter is necessary for effective modern quality management. It is hoped that the outline of quality education syllabi provided will be of some help.

It should be remembered that the present thinking and attitude of employees was developed over a long period of time and changing this in a positive way so that it never slips back will require ongoing quality education with refresher courses given on a regular basis.

Quality education is a key factor in the success of quality management in any organisation. It must be structured and planned, as well as continuous.

7.6 Internal Quality Auditing

Internal quality auditing is the least costly of the processes an organisation can use for continually identifying risks and driving improvement, and when supported and encouraged by top management, is also the most powerful.

Internal quality audits, as a requirement of ISO 9001 the International Standard for Quality Management System Requirements and GB/T 19001, are necessary to determine whether the organisation's quality management system conforms to the requirements of the Standard, to the quality management requirements established by the organisation, and that the organisation's quality management system is effectively implemented and maintained.

Internal quality auditing provides a means to verify the effectiveness of application and control of the quality management system processes, and to assess strengths and weaknesses, and therefore to identify actions needed to improve process and system effectiveness. An effective internal audit process will uncover risks associated with the lack of control, and verify that corrective and improvement actions are effective and performed in a timely manner. In addition, internal quality audits will enable management to obtain a measure of its own effectiveness in controlling the operation of the organisation in the manner intended.

The better Western organisations take an approach that allows quality auditing to be easy to manage and cause minimal disruption to operations; they plan highly focused and short duration audits of the various processes over an entire year. An "Internal Audit Schedule" will be developed at the beginning of each year indicating what will be audited and during which week in the year each targeted process will be audited. This is a "best practice" approach to internal quality auditing.

The internal audit process is an ongoing activity, continually auditing different parts of the system and covering all areas of the organisation. There could be two or more of these audits of different processes of the organisation spread out over a month. Areas where the audit findings are adverse receive more frequent audits until such time as the audit results stabilise; particularly important activities are also audited more frequently.

Audits of conformity with documented procedures ensure that processes remain stable – importantly, with the audit comes the review of operating practices and results; this enables an organisation to constantly find new and better ways of doing things which can then be formally incorporated into the way of working.

Perhaps the worst approach is the scheduling of an audit of the entire quality management system just once a year. This has been seen to take place in organisations in China; this audit of the entire quality management system usually occurred over a one or two week period, just before the annual external audit by a GB/T 19001 (ISO 9001) national accreditation body.

There is so much ineffectiveness with this practice and it suggests that the true intent of the internal quality audit process is misunderstood. The once a year practice makes it impossible to secure any momentum of benefit, the practice does not promote a growth mind-set in the organisation, and trained internal quality auditors soon become rusty as they are only able to practise the skill of auditing for a brief period in the year. Furthermore, it causes disruptions to operations in the organisation and places excessive strain on the auditors and audited parties. The scope is so wide that auditors tend not to delve too deeply so non-compliances are easily overlooked and improvements are not identified. Furthermore, it is not possible for top management to conduct an effective management review of the quality management system based on the results of such an audit.

With qualified quality personnel guiding the quality management effort, the selection of what and when to audit is based upon status and importance; this is an essential element of "best practice" internal quality auditing. An audit of the entire quality management system, done once a year over a one or two week period, gives virtually no consideration to the meaning of status and of importance.

Status concerns how a particular unit, department or process is performing against expectations and established goals and objectives, and is reflected in, for instance, its quality performance history, changes in process or equipment or key personnel, and re-structuring or re-organisation of departments.

The essence of the meaning of importance is with regard to the power to produce an effect, e.g., the importance of the inspection process lies in its effectiveness in preventing non-conforming products from being dispatched to the customer. Importance is directly linked to status, for instance, when a status change occurs by way of re-structuring a department, or when facilities are relocated, time is usually a major concern to get things operating as soon as possible. In such cases it is the auditing of the implementation and functionality of the re-structured department or relocated facilities that become important.

Effective audit schedules do not only contain audits with a focus on obvious shortcomings or weaknesses in an organisation, they also contain audits focused on areas that could potentially be a cause for non-conforming or risk situations, as well as areas critical for maintaining product conformity.

Table 7-6 contains examples of problems, shortcomings or concerns that were found in organisations in China and the associated processes that were identified as needing to be audited. The issues identified risks to the organisation and most of these incurred PONC which robs profits.

Table 7-6: Examples of issues and processes identified for quality auditing

Problem, Shortcoming, Concern	Processes to be Audited
Shortcomings in achieving quality objectives	Planning and implementation of quality objectives (in the various functions and departments)
Many late deliveries	Begin here: Customer order acceptance (including contract and technical review and order scheduling)
Defects in product received by the customer	- Final product inspection and testing - Operator Self-Control
Gearbox oil leaks occurring in the warranty period	Design review of the gearbox (with emphasis on design actions to eliminate oil leaks)
Inadequate use of information obtained from component inspection results	- Collection and analysis of quality data (including use of data analysis techniques) - Continual improvement and preventive action
Inadequacies in the customer complaint handling process associated with response to closing technical problems	Handling of technical customer complaints (from registration through to closure, and the competency of personnel involved)
Failures of conveyor chain found upon final inspection and test (e.g., low breaking force, excessive weld trim)	Validation of processes, and monitoring and measurement of product in the Chain Factory
Unauthorised use of non-conforming product and lengthy time taken to disposition non-conforming product	Control of non-conforming components and product
Damage due to bumping (collision) and corrosion of components	Identification, handling, packing and storage of components
Move to the new premises	Provision of infrastructure and suitable work environment, and control of records

Internal quality auditing of associated processes will uncover the reasons for the issue and indicate where corrective action is required, or uncover problems that need to be rectified.

Many of the processes identified as needing to be audited span different functions or departments, and their effectiveness can only be determined when all things that shape and support the overall process are taken into account. The audit for process effectiveness must therefore include interface activities with other processes and take into account the requirements of internal suppliers and internal customers, i.e., the quality audit must take a process-based approach.

In some organisations in China, an internal quality auditing approach was seen where the internal auditors concentrated on a particular function or department, and focused on the use of procedures used in the function or department and the compliance of the procedures with relevant clauses in GB/T 19001 (ISO 9001). This approach is generally limited in effectiveness especially as it could support self-serving activities and be blind to the requirements of internal customers.

If a process-based approach is not adopted when conducting quality audits, it is likely that the root cause of the problem or shortcoming will go undetected.

The audit schedule is a living document that can be revised through the year as may be required to take into account changes and findings from such things as management review outputs, customer concerns, internal quality issues, supplier and sub-contractor quality problems and organisational infrastructure changes.

Note that in response to demand for guidance on combined management system audits, ISO 19011 (GB/T 19011) was published. This is an instructive standard providing guidelines for management systems auditing, notably quality and environmental as well as safety management systems; ISO 19011 is explained in Appendix 1.

Questions for Chapter 7

7-1: Identify at least six reasons that could bring about the need for Quality Planning in an organisation.

7-2: Provide several main purposes of Design Reviews.

7-3: What are some of the shortcomings and risks inherent in relying on end-of-process inspections by an independent inspection function?

7-4: Outline the general steps concerning the collection and analysis of data for use as quality information.

7-5: Why is quality education important for top and middle management?

7-6: What are the purposes and aims of internal quality auditing?

7-7: Outline what needs to be considered in an effective internal quality audit program.

PART 4: ENABLING, SUPPORTING & IMPROVING PERFORMANCE

Subjects, practices, concepts and techniques contained in Part 4

Chapter 8: Operational effectiveness
- ♦ The role of core and support processes in value creation
- ♦ Operation of the Work Process Model
- ♦ Topical practices relating to operational performance
  - Piece-work
  - PDCA, Plan–Do–Check–Act
  - OEE, Overall Equipment Effectiveness
  - PONC, Price of Non-Conformance
  - Lean Thinking
  - 7 Wastes
  - 5S/5C

Chapter 9: Supplier quality assurance
- ♦ Supplier selection – supplier questionnaire, supplier quality assessment, supplier classification
- ♦ Ensuring correct supply – quality plan for bought in product, receiving inspection, surveillance visits, quality visit reports, supplier corrective action request
- ♦ Supplier review – supplier quality history record, supplier rating
- ♦ Supplier development – supplier development programmes

8. OPERATIONAL EFFECTIVENESS

Top management in leading organisations is continually seeking out strategies that will give their company a competitive edge. This is driven by the fact that competitive advantage ensures their company's prominent placing, respect and growth in the marketplace.

The combination of a good strategy and operational effectiveness is necessary for business success. The strategy chosen may be one of, or a combination of strategies such as cost leadership, product differentiation, innovation, or niche-market, but, for the strategy to succeed, operational effectiveness is most necessary.

The activities discussed in this chapter for better operational effectiveness include:

1. The arrangement and information exchange between processes
2. Enabling, supporting and improving operational performance
3. An effective "Price of Non-Conformance" reduction programme
4. Maximizing value whilst minimizing waste through "Lean Thinking"

8.1 The arrangement of processes and information exchange

The organisation's core competencies are contained in functions and processes, and for operational effectiveness in the implementation of the organisation's chosen strategy, they need to work well together.

Manufacturing organisations commonly organise themselves into vertically functioning groups (e.g., sales, design, production, customer service) in which people with knowledge and skills are brought together and trained to be capable of completing tasks within that function. This creates strong functions; however, there are two prime issues that need to be addressed:

- Firstly, most processes flow horizontally through the organisation, across vertically functioning groups, and many voids can result which negatively impact the efficiency and effectiveness of the processes.

- Secondly, often the function can be inward-looking and eager to support its own goals or mission that may not be totally aligned with the overall needs of the business.

Most organisations tend to retain the functional organisation as it provides benefits to core competency by bringing together like-minded people in functions, but the best organisations think in terms of value creation through processes.

Value creation:

Processes in an organisation are either "core" or "support" processes; core processes are the value creators; support processes provide the core processes with essential help so they can create value. Core processes are also called "business" or "primary" processes.

Shown at the centre of figure 8-1 are the high-level "core or business processes" for an organisation that develops products, markets them, acquires customer orders, produces, delivers, installs and services its products. These processes contain competencies that are unique to the organisation and that are critical to the purpose of the organisation.

Support processes are depicted in figure 8-1 as "management-related" and "service-related".

The management-related support processes provide direction, strategy and plans for the organisation, and they provide management systems and manage resources and the attainment of plans and goals; they shape and manage the business and support processes used by the organisation. Quality management processes are an intrinsic part of the management-related support processes.

The service-related support processes exist to sustain the business.

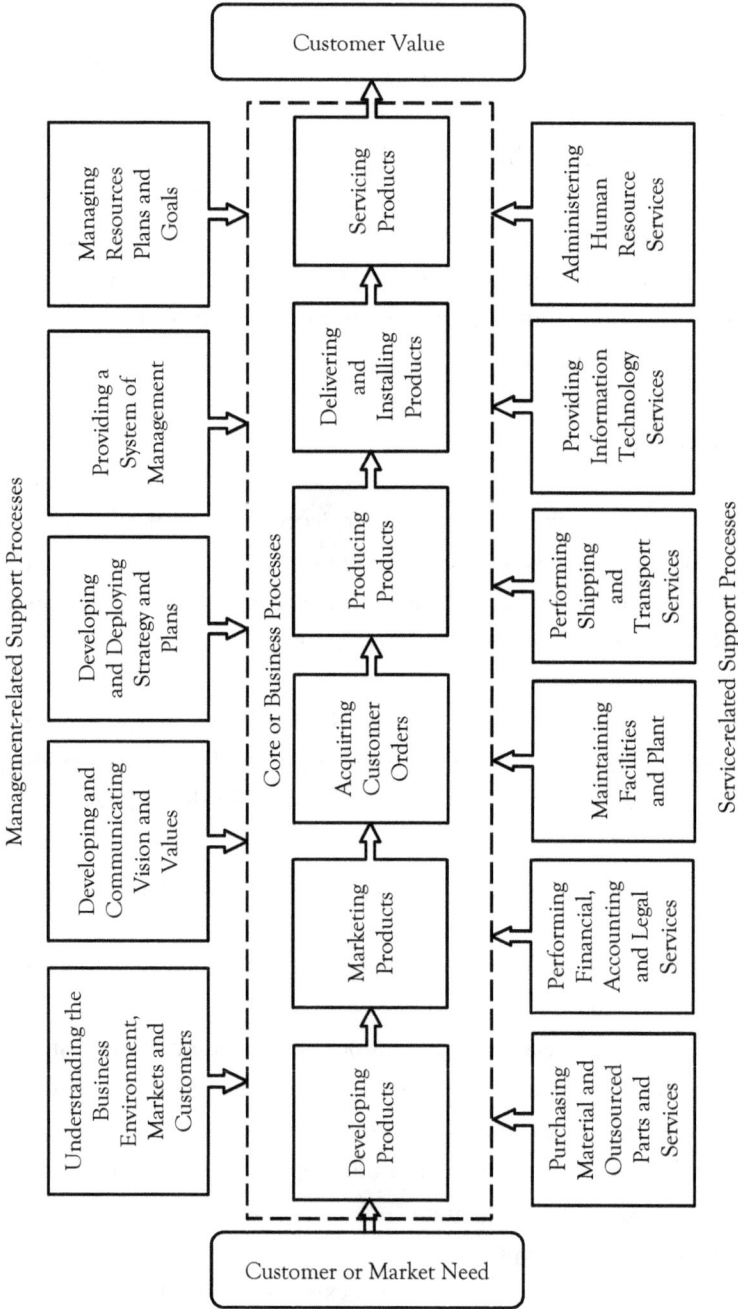

Figure 8-1: Core and Support processes

The model in figure 8-1 embodies a process approach. It illustrates that the output of each core process serves as an input to the next core process. The downstream process is the internal provider or supplier to the upstream process. The upstream process is the internal recipient or customer for the downstream process. Through efficient and effective process interaction, and with the help of support processes, the driving intent is to satisfy customer or market needs, and ultimately create value for the end-customer and generate overall success for the organisation.

Distractions from value creation are observed to occur when;

- Service processes place constraints on value creation processes; for instance, in establishing manpower requirements for the production process and HR dictates how many production operators will be allowed. In contrast in the West, Production will establish manpower requirements based upon their knowledge, experience, and work-study information, and then they would ask HR to provide the service of recruiting to these requirements.
- Functions concentrate on their own mission, or are obsessed with their own importance and do not grasp that the overall organisation's success suffers as a result.
- The organisation is burdened with bureaucracy. Bureaucratic organisations have rigid procedures with stringent controls; they operate in a formal manner and are inflexible and reluctant to change. There are many layers of management; a strict command and control structure is present at all times and decisions are made through an organised process. In a business where the focus is on value creation, bureaucracy is harmful as it slows down transactions between the many functions and processes of the organisation. Bureaucracy is purposefully avoided in the better business organisations.

Operational efficiency in the work process:

Within each of the high-level processes in an organisation are numerous work processes. If the output of the downstream process fails to meet the defined input requirements of the upstream work process, this has a negative effect on the performance of the upstream work process, which could be the supplier to the next internal customer or the end-customer. Figure 8-2 illustrates a work process model.

In this model, the suppliers, which could be internal or external, provide the inputs that meet defined requirements. This is necessary in order for the work process to be able to produce outputs that meet the defined output requirements. Also, when the internal supplier knows the requirements of its internal customer it is able to identify and control the delivery of its own input requirements.

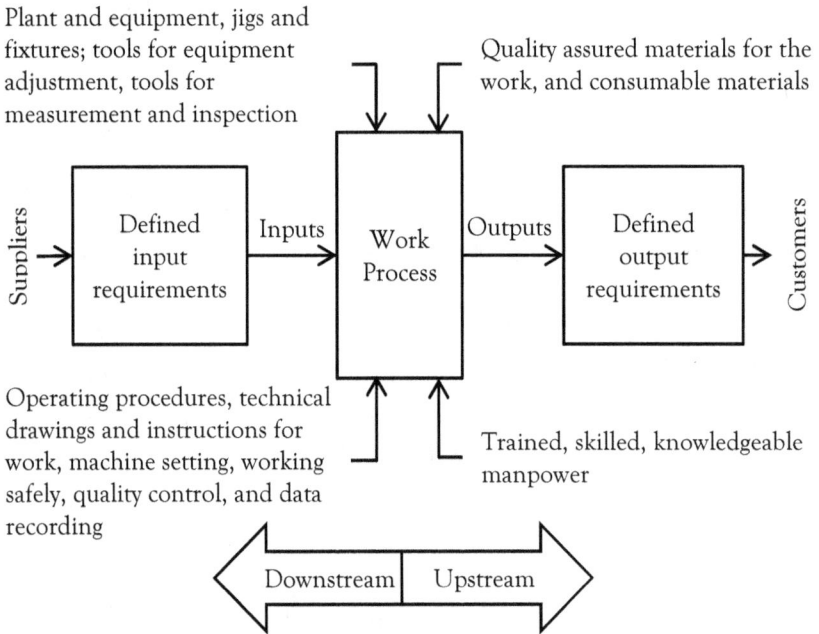

Plant and equipment, jigs and fixtures; tools for equipment adjustment, tools for measurement and inspection

Quality assured materials for the work, and consumable materials

| Suppliers | → | Defined input requirements | Inputs | Work Process | Outputs | Defined output requirements | → | Customers |

Operating procedures, technical drawings and instructions for work, machine setting, working safely, quality control, and data recording

Trained, skilled, knowledgeable manpower

Downstream | Upstream

Figure 8-2: Work Process Model

The internal customer concept is very important however employees must not lose sight of the fact that the focus of the organisation is on creating value for the end-customer; absolute obedience to the requirements of internal customers may not be advisable or operationally efficient for the performance results of the organisation – for instance, the end-customer would not be happy if he is made to wait, or made to pay for some or other organisation borne bureaucratic requirement to be fulfilled. From the outset, a mutually agreeable set of requirements that are ultimately based on the requirements of the end-customer must be agreed through dialogue between internal customers and suppliers.

8.2 Enabling, supporting and improving operational performance

Among key activities taken by better organisations to enable, support and continuously improve operational performance, are the following:

- Each process in the process-chain has its output requirements clearly defined, and for each process output, checks are carried out to ensure that its internal customer requirements are met.
- The configuration of production lines is analysed and optimally balanced for the most practically efficient flow that avoids build-up of work-in-progress and excessive handling of material and product. This is because excessive handling is not only motion-wasteful, but it increases the potential of product being damaged.
- The layout of each workstation, regarding the movement of material and the operator in his workstation, is optimised to create order and a safe environment, to minimize movement and to enable quality control actions (inspection, identification, recording, segregation).
- Keen interest is taken in the competency and performance of work performers. Operators are continually evaluated, trained and educated. Training includes on-the-job skills training, and in quality control actions specific to the work. Education is provided in quality awareness, health and safety and working safety, environmental practices, and as the need arises, in application of quality tools.
- A Local Area Network is set up to provide operators access to approved and controlled information. This not only gives easy access from within the work area to needed information, such as process procedures, technical information, and quality control instructions, but also prevents the use of obsolete documentation.

- Work progress is tracked on computer systems through all phases of manufacturing and quality control – Material Resource Planning (MRP II) is an example of an information system that has this functionality. When practical, bar-coding is often employed to track the progress of work from one workstation to the next.
- In some organisations there are help-call buzzers at workstations which the operators are instructed to use to alert the supervisor of, for example, material and machine problems, and to alert the quality control technician of quality issues.
- Measurement and information collection systems are employed to report types and number of defects, to indicate process variability, and to report the overall utilization of facilities; this information is used to promote and advance continuous improvement.

Over the years companies have tried various means and methods to improve or enhance operational effectiveness. Some have stood the test of time and been further developed, others have not been successful and have lost favour. Discussions follow in this chapter concerning;

(1) Piece-work – these schemes, thought to promote productivity, are still popular in industrial manufacturing organisations in China
(2) PDCA method – this is a very powerful method for making changes or implementing new ideas in a controlled manner
(3) OEE measurement – this is a good basis which is frequently used to evaluate and improve the Overall Equipment Effectiveness of a particular manufacturing operation

(1) Piece-work:

Management in industrial manufacturing organisations in China want high productivity and, to encourage high labour output, piece-work schemes are commonly used where the operator is paid for good pieces that he makes and penalised for non-conforming and scrap items. He is expected to inspect his work and to declare all defects that he finds. This leads to the tendency for operators not to be motivated to check their work, and many operators hide non-conforming and scrap pieces.

Piece-work schemes were tried in manufacturing organisations in the West up until the 1980s. Attempt after attempt failed in realising the intent of piece-work schemes to improve productivity, they therefore became unpopular. Reasons why they lost favour include the following:

- The incentive of any piece-work scheme is to meet and exceed the production target, not to satisfy the customer. The worker's main concern is about the quantity of work completed and not the quality. The operator, being paid according to the number of parts he produces, would rather allow a defective part to pass than to inflict self-penalization.

- Industrial engineers spent a lot of time trying to overcome the negative effect of the operator penalizing himself when he declared non-conforming parts, and introduced quality grading; grading of quality did not encourage good quality.

- The pieces or parts produced by the operator had to meet certain criteria to be classed as in-conformance and close examination by another party was performed; this fuelled friction.

- The task of determining piece-work targets encouraged gamesmanship and actually promoted low expectations; when the operator saw the industrial engineer approach with the stopwatch, the operator reasoned that it made more sense to conceal how much he could do, so he deliberately slowed down his work rate.

- The scheme discouraged suggestions from operators to improve productivity; the tendency was for innovative operators not to share their "secret techniques" that enabled them to beat the set quota.

- Piece-work schemes applied to individuals; this discouraged them from spending time in worker-to-worker participation for better quality, e.g., feedback on quality issues found on parts supplied by their feeding cell or internal supplier.

- It was found difficult to measure a person's overall value-adding benefit based solely on the units they could produce in a given period of time.

- Piece-work schemes required a lot of costly administration that became difficult to justify.

The better Western management recognises and favours the strength of the benefits arising from worker supporting worker, and the power of teamwork with clear focus on meeting customer requirements, and profit-related incentive schemes leading to better productivity.

(2) PDCA method:

The PDCA method is used in making improvements and finding optimum solutions; it has four phases and is depicted in figure 8-3.

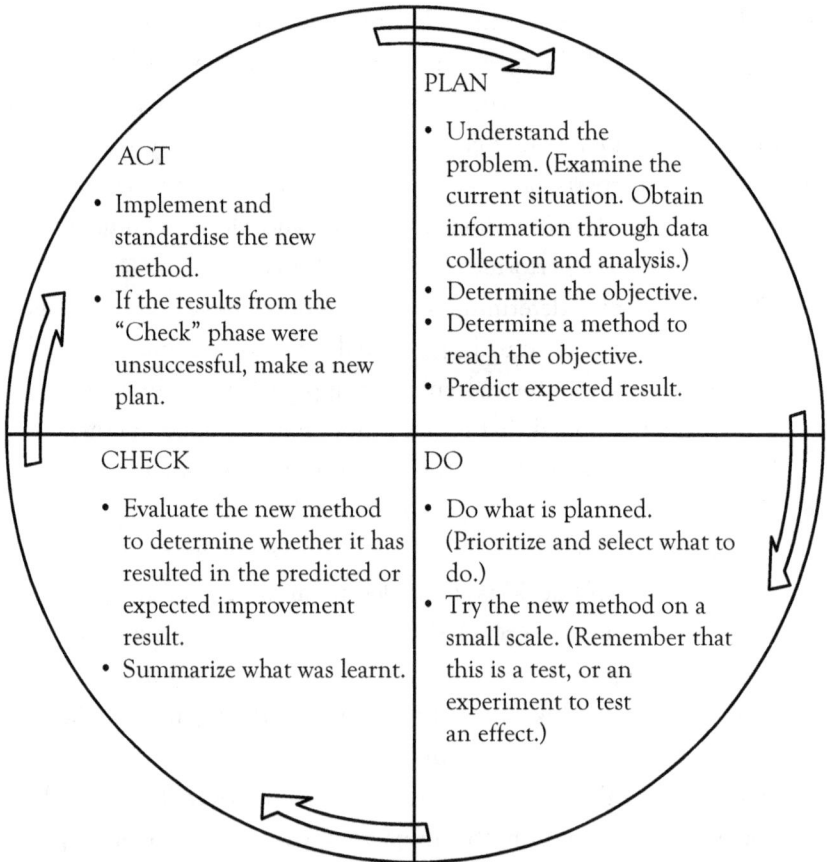

Figure 8-3: PDCA

PLAN
- Understand the problem. (Examine the current situation. Obtain information through data collection and analysis.)
- Determine the objective.
- Determine a method to reach the objective.
- Predict expected result.

ACT
- Implement and standardise the new method.
- If the results from the "Check" phase were unsuccessful, make a new plan.

CHECK
- Evaluate the new method to determine whether it has resulted in the predicted or expected improvement result.
- Summarize what was learnt.

DO
- Do what is planned. (Prioritize and select what to do.)
- Try the new method on a small scale. (Remember that this is a test, or an experiment to test an effect.)

The "Plan" phase begins with a good understanding of the problem; the basic requirements for making a plan are summarized in the first quadrant of figure 8-3.

When a plan has been agreed, the planned actions are implemented in the "Do" phase on a small scale. It is important to be clear that "Do" means to *try* or *test*, it does not mean to *implement fully*; implementation happens later in the "Act" phase.

"Check" occurs in the next phase; the effects of the planned actions are evaluated. If the expected or predicted results are produced, the change is accepted and implemented in the "Act" phase, if the expected results are not produced, or if further improvements are required, a new plan is made and the cycle is repeated.

PDCA is more than a method for making changes or implementing new ideas in a controlled way, it is a way of thinking. The method encourages thoughts of what could be achieved. It also removes impediments to trying, and frees individuals inhibited by the thought of losing face. PDCA is a learning method based on experiments. Writing down the predicted results forces more careful up-front thinking, helps to set better experiments, and helps the learning process.

When used properly with management support, the PDCA method can improve quality and productivity.

The following narrative, concerning excessive burrs on edges of the machined surfaces of gearbox housings, demonstrates the power of the PDCA method in improving production efficiency and effectiveness. This occurred in an industrial organisation in China.

After machining of gearbox housings, edges of the machined surfaces exhibited burrs. Burrs were unacceptable, regardless of their degree, and were required to be removed.

Burrs occurred to various degrees; sometimes they were minimal and took minutes to remove, at other times they were excessive and took a lot longer to remove. The de-burr operation took several times longer than the machining operation and there was insufficient floor space near the machining workstation on which to store the housings waiting to be de-burred. The machined gearbox housings were therefore transported across the workshop to an area where there was space. This became the designated de-burr workstation. This process of machining at one workstation and de-burring at another workstation was the accepted norm of doing the job. The bump-damage that occurred in the handling process often resulted in further rework.

It was agreed with senior management that the process of machining gearbox housings could be examined with the view to improving the situation, and a non-disruptive PDCA method would be employed during normal production where various changes would be tried.

Using a Cause and Effect Diagram (refer to Chapter 11.5), it was shown that burrs could be associated with a number of variables; in this case they were associated with the cutting parameters, the cutting tool, the rotational speed of the machine, and the machining lubricant. The effect of each variable was investigated using a Scatter Diagram (refer Chapter 11.6) to establish correlations.

The PDCA method helped to resolve an optimum situation in which cutting speed, feed rate, cutting tool material and shape, frequency of changing the cutting tool, and machine rotational speed were re-defined. The effect on burr-reduction was dramatic. The de-burr operation now took less time to do than the machining operation; there was no need to temporarily store an accumulation of housings waiting de-burr. This made it possible for a small de-burr workstation to be included in a slightly enlarged machining operation area.

A few simple steps had identified optimum machining conditions which had negligible effect on the piece-work times, the throughput constraint caused by the previously lengthy de-burr operation was removed, and the incidence of bump-damage was drastically reduced.

Failure to continually work at improving operational effectiveness throughout the organisation allows dangerous sub-optimal performance to become accepted as the way of life. The PDCA method can be used to challenge long-standing sub-optimal practices and attitudes of resignation to a given situation.

The PDCA idea underpins KAIZEN, which works to optimise and improve existing process controls, and the Six Sigma project methodologies. KAIZEN and Six Sigma are explained in Appendix 1.

(3) Overall Equipment Effectiveness (OEE) measurement:

OEE measurement is frequently employed to evaluate how effectively a specific manufacturing operation is utilized. The manufacturing operation could involve that of a particular workstation or a distinct piece of equipment. OEE measurement indicates the gap between actual and optimally ideal performance such that improvement action can be identified and taken. OEE measurement and the individual OEE factors of Availability, Performance and Quality are powerful performance indicators that are intended to be compared to each other on a production run-to-run basis of a specific manufacturing operation.

The individual OEE factors and OEE measurement are defined and calculated as follows:

Availability:

"Availability", for the purpose of OEE is the actual time in which the manufacturing operation is available to operate as a percentage of its planned time. It is a measure of uptime and excludes downtime events.

(1) *Availability % = (actual operating time /planned operating time) × 100*
(Note: planned break time (e.g., operator rest-break time) must be excluded when calculating planned operating time)

Performance:

"Performance", for the purpose of OEE is the net operating time of the manufacturing operation as a percentage of its actual operating time. It excludes the effects of Quality and Availability.

(2) *Performance % = (net operating time / actual operating time) × 100*
- net operating time = optimal cycle time × total units started
  (Note: "total units started" excludes the effect of Quality)
- cycle time = time/unit; rate (reciprocal of time) = unit/time
- optimal rate = total count / net operating time

- actual rate = total count / actual operating time

Substituting from the above,

(2.1) *Performance % = (optimal cycle time × total units started / actual operating time) × 100*

(2.2) *Performance % = (actual rate / optimal rate) × 100*

Quality:

"Quality", for the purpose of OEE is a measure of process yield of the manufacturing operation, also referred to as "First Pass Yield", i.e., it is a measure of the units that successfully pass through the manufacturing process the first time without needing any rework. It is designed to exclude the effects of Availability and Performance.

(3) *Quality % = (good units produced / total units started) × 100*

OEE measurement:

This is the product of Availability, Performance and Quality.

(4) *OEE % = Availability % × Performance % × Quality %*

Example of OEE improvement:

The Process Engineering Department calculates that an optimal production rate at a particular workstation is 30 units per hour, i.e., a cycle time of 2 minutes/unit. The workstation shift duration is 9 hours and a total of 60 minutes is scheduled for operator rest-breaks.

Before improvement actions, the workstation produces or starts the production of a total of 180 units during the 9 hour shift. In the 9 hour shift, 7 of these units have quality problems resulting in a good product output of 173 units. During the shift duration there are 76 minutes of unscheduled downtime.

Availability calculation:

Planned operating time is workstation shift time less the scheduled rest-break time: planned operating time = (9 × 60) – 60 = 480 mins

Actual operating time is planned operating time less unscheduled downtime; actual operating time = 480 - 76 = 404 mins

Availability % = (404/480) × 100 = 84.2%   (1)

An Availability of 84.2% implies that the workstation was not available to produce for 15.8% of the planned operating time. Efforts should be directed at determining the causes of downtime losses and reducing them. Downtime could be caused by waiting for material, tool breakage, frequent adjustment of the machine, or machine problems of a hydraulic, mechanical or electrical nature.

Performance calculation:

Optimal production cycle time for the part is 2 minutes/unit.

Net operating time of the manufacturing operation is the optimal production cycle time multiplied by the total units started.

Net operating time = 2 mins/unit × 180 units = 360 mins

Actual operating time = 404 mins

Performance % = (360/404) × 100 = 89.1%   (2.1)

A Performance of 89.1% implies a production speed loss of 10.9%. Efforts should be directed at ascertaining the causes of the constraint. This may be due to machine loading or material handling difficulties, a slower machine cycle time setting, or an untrained operator.

Quality calculation:

Good units produced right the first time = 173

Total units started = 180

Quality % = (173/180) × 100 = 96.1%   (3)

A Quality metric, or First Pass Yield, of 96.1% implies a Process Yield Loss of 3.9%. Efforts should be directed at finding the cause(s) of the Quality issue(s). Figure 2-1, provides a general checklist of possible causes of poor quality.

OEE measurement before improvement actions (OEE_1):

OEE_1 % = 84.2% × 89.1% × 96.1%   (4)

Therefore OEE_1 % = 72.1%

Over a period of 8 weeks actions were taken to improve Availability, Performance and Quality, and therefore the OEE.

Availability was found to be affected by the operator having to frequently adjust the machine and by overlooked maintenance issues. A Control Chart was used to study machine variability and to optimise machine adjustment intervals. The 5S/5C "Red Tag" system highlighted many temporary fixes applied to keep the machine running that were collectively causing machine use restrictions. Corrective actions were taken to eliminate the "Red Tags". Both actions resulted in a reduction of unscheduled downtime from 76 minutes to 21 minutes. The Availability % increased to 95.6%.

The operator was given training which reduced machine loading time. This, together with the increase of uptime to 459 minutes, resulted in more time to produce – in the 9 hour shift, the workstation loaded 209 units. The Performance % went up to 91.1%.

Following a study of the machine variability, the operator was given instructions on how frequently to conduct in-process measurement and apply adjustments to the machine to prevent non-conformances. In the 9 hour shift, only 3 of the 209 units had quality problems resulting in a good product output of 206 units and a Quality % of 98.6%.

OEE measurement after improvement actions (OEE_2):
$$OEE_2 \% = 95.6\% \times 91.1\% \times 98.6\% \qquad (4)$$
Therefore $OEE_2 \% = 85.9\%$

Discussion of the overall result:
In 8 weeks, improvement in each of the OEE factors caused an overall improvement of OEE from 72.1% to 85.9%.

OEE measurement expressed in a single number provides a succinct way of reporting to top management how effectively a particular manufacturing operation is utilized. However it needs to be borne in mind that a single OEE number does not provide an insight into the nature of loss; the three factors of Availability, Performance and Quality draw out the underlying issues caused by various types of waste such as equipment down time, changeover time, machine-loading time, slow cycling time, and quality problems.

For instance, OEE_2 of 85.9% seems to be very good and certainly better than the previously obtained OEE_1 of 72.1%, yet consider if OEE_2 was achieved from an Availability of 99.5%, Performance of 95.9% and Quality of 90.0%. It is hard to imagine that any company would agree that it would be acceptable to increase Availability and Performance at the expense of Quality, which, in the case in question is a decrease from 96.1% to 90.0%.

OEE measurement in industrial manufacturing organisations in China has a place in promoting improvement in the effective utilization of both old and new equipment, especially of costly equipment where return on investment is important.

8.3 "Price of Non-Conformance" reduction programme

Quality Costs can comprise a major portion of the total expenses of a business; unfortunately they are not disclosed in the cost accounting system which is oriented towards recording by responsibility centre and not by quality issue. Identifying Quality Costs builds awareness of the importance of quality and identifies opportunities for improvement.

Quality Cost reports have indicated that changes need to be made when Appraisal and Failure Costs are relatively high and Prevention Costs are low; this indicates a need to increase spending on prevention activities.

The results of Price of Non-Conformance (PONC) surveys have never ceased to be eye-openers for management. The shock to top management has occurred when findings have established PONC to be in the region of 20% to 30% of Profit Before Interest and Tax (PBIT). In each case the finding, verified by the accounting department, has prompted top management to take action to reduce this waste. It should be stressed that it is not uncommon for the PONC survey to uncover PONC of 20% to 30% of PBIT; higher percentages of up to 45% have been found.

When examining the price of poor quality, the "Iceberg Principal" comes to the fore, i.e., the obvious price of poor quality from rework, scrap and warranty, is readily visible. Figure 8-4 illustrates the "Iceberg Principal".

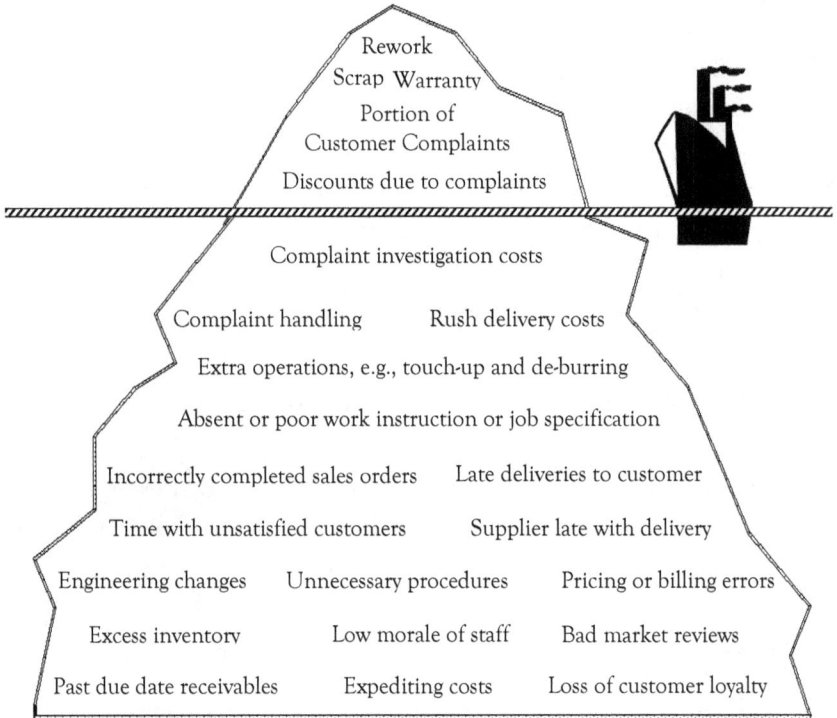

Figure 8-4: Cost of poor quality – The Iceberg

In Western organisations the readily visible portion of the iceberg typically ranges from 4% to 8% of sales.

It was found that industrial manufacturing organisations in China report some of the contributors to PONC, notably obvious waste seen at the tip of the iceberg occurring from scrap and product failure in the field, however, scarce attention was given to hidden or indirect PONC, i.e., that which is below the iceberg's waterline. Findings indicated that hidden PONC was caused typically by;

- "Over-promising" by sales-personnel, as well as inadequate application of the contract review process, where the organisation should have checked that it could do the job to the requirements specified by the customer. This often resulted in failure to achieve customers' requirements, and late deliveries.

- Lack of quality control in design resulting in the likes of producibility problems due to tolerance issues, and extra-operations such as re-cutting of assembly holes.
- Lack of quality control at workstations by not performing regular in-process checks.
- Non-conformances found at the end-of-process inspection, i.e., late in the production cycle, carried the concern of high value-added waste. Much of the non-conforming material or product was inevitably used (passed to further processes down the line) because engineering usually conceded, in their material disposition decision, to a generous increase in acceptance limits. The lengthy processing of such material disposition applications increased the volume of work-in-progress and contributed to production delays. The extra handling resulting from moving the material into quarantine storage incurred further costs and often resulted in bump damage.
- Errors made when preparing the invoice and/or preparing the product for dispatch such as incorrect quantities and incorrect delivery address.
- Non-conformances, deficiencies and product failure during product service life.
- Giving discounts to customers to buy back goodwill.
- Possible loss of future business triggered by long delivery delays. Delivery delays were often several months in duration.

At the bottom of the iceberg in figure 8-4 is the cost incurred through expediting and past due date receivables. These are typical of the costs that are regarded by many manufacturers as normal business expenses but do in fact contribute significantly to PONC.

"Expediting costs" are the monetary expenses of the purchaser's resources accrued in the actions taken to hurry attainment of a required outcome. Practically all expediting costs are as a result of failure to control quality.

In Western organisations, particularly up to the mid-1990s, it was not uncommon to find a large number of dedicated purchasing expeditors chasing renegade suppliers – e.g., suppliers late in delivering goods, or delivering incomplete orders, or delivering orders that did not conform to requirements.

Purchasers came to realise that they were wasting their resources and harming themselves by allowing suppliers to cause them to miss end-customer delivery dates. Penalty clauses could not be applied successfully to all suppliers, so, through careful evaluation, selection and the application of supplier development programmes, purchasers gradually reduced the number of renegade suppliers. (Details of supplier selection and development are contained in Chapter 9.) The need for purchasing expeditors reduced – nowadays, in the West, it is rather unusual to find many employees dedicated to the job of expediting.

In some industrial manufacturing organisations in China, supplier/sub-contractor selection procedures, vendor agreements, and even supplier technical support and supplier/sub-contractor rating systems were in evidence; these are all good elements for effective supplier management, however observations and findings indicated that many times their application was perhaps a little inconsistent and sometimes not entirely proactive and decisive. Generally there were a large number of purchasing expeditors employed chasing late deliveries and sorting poorly conforming bought-in goods; this added to the PONC in these organisations.

Past due date receivables are payments on Accounts Receivable for products and services that are not made within a specified time. This obviously is not good for the seller's cash flow and can cause the seller to incur overdraft expenses or worse.

Poor quality management, resulting in failure to control quality, can cause many invoices not to be paid, or not to be paid in full. Some of the common reasons for delayed payment due to poor quality control include;

- disputes regarding conformance or quality of work performed;
- invoicing errors, e.g., incorrect particulars or an inaccurate invoice statement;
- incomplete delivery, e.g., incorrect number of items or a missing technical test report

These are valid reasons for the purchaser to withhold payment and are therefore targeted to be eliminated by way of quality control measures applied to both the support and core company processes.

8.4 Maximizing value, minimizing waste through Lean Thinking

"Lean Thinking" sets a pattern of thought and approach to work that is guided by the following broad principles:

(a) The expenditure of resources for any goal other than the creation of value for the end customer is regarded as wasteful and thus targeted for elimination.

(b) The implementation of smooth flow exposes constraints and quality problems, thus action is directed towards overcoming these issues.

Quality management establishes a suitable environment for "Lean Thinking" which facilitates operational effectiveness; the PONC Reduction programme discussed in Chapter 8.3 directly supports (a) above, and the process approach discussed in Chapter 8.1 supports (b).

Management in progressive organisations in the West began taking an interest in Lean Manufacturing in the 1990s. Initially a lot of emphasis was placed on the use of Lean tools or methodologies such as Value Stream Mapping, 5S/5C, Kanban (pull systems), poka-yoke (error-proofing), Total Productive Maintenance, charts for monitoring excessive variability, and SMED (Single-Minute Exchange of Die – the rapid changeover of tooling).

Lean tools were found to individually bring benefits, but sustained benefits resulted when organisations adopted "Lean Thinking" which influenced behaviour.

The "Lean Thinking" targets became the elimination of wasteful expenditure (PONC), elimination of wasteful motion and non-value-adding actions, and implementation of smooth process flows.

The agenda in the better organisations includes quality improvement, PDCA and the elimination of the "7 Wastes".

7 Wastes:

In industrial manufacturing organisations in China it is not uncommon to see posters proclaiming Lean Manufacturing, but the evidence points to another scenario: high levels of work-in-progress on the production floor, stores full of non-moving stock, a lot of product handling and transport around manufacturing areas and the site, tolerances that are set excessively tight, and often employees pretending to be busy.

"Lean Thinking" in the organisation begins with the understanding of the "7 Wastes" of Lean Manufacturing; these are as follows:

1. Over-processing / over-complication (in all departments):
   This is caused, for example, where operators are required to work to tolerances that are too tight, use inappropriate or non-optimum techniques or equipment, are not given clear acceptance limits (e.g., for surface finishes), and where employees are required to go through a long procedure (e.g., to get a decision on the disposition of non-conforming product). Processing-type wastes can also occur when customer specifications are unclear, or there are frequent engineering changes, and where inadequate value analysis has been performed.

2. Idle time / waiting (in all departments):
   This is the waste caused by the act of doing nothing or working slowly whilst waiting for material or the completion of a previous

step in the process, or by an employee waiting for his team leader to give him the next task, or waiting for a delivery from a supplier, or waiting for an engineer to come and fix a machine. Waiting can also be caused by unsynchronized processes, line imbalance, an inflexible work force, long set-up times, material and manpower shortages, and unscheduled machine downtime.

3. Bad quality (in all departments):
   Bad quality or defects in products cause customer dissatisfaction and possible damage to company image, and waste of resources and waste of time caused by rework, repair, replacement, or making good. Preventive measures and continuous quality improvement are the most effective means to cut bad quality wastes.

4. Over-production:
   The waste of over-production is making too much or producing too early. This could be caused by production volume-based incentives (common in China), high capacity equipment, line imbalance, poor scheduling or poor production planning, and cost accounting practices that result in a build-up of inventory. Over-production leads to high levels of inventory which mask many of the problems within the organisation. The aim should be to make only what is required when it is required.

5. Excessive inventory:
   Every part of the product tied up in raw material, work in progress or finished goods, has a cost and until the product is actually sold that cost is the burden of the organisation. Furthermore, excessive inventory takes up storage space and product held in storage for a long time can deteriorate or become obsolete. Excessive inventory also results in increased handling, and product has the chance of being damaged during handling.

6. Unnecessary movement and transport:
   Moving and transporting is a waste as it adds no value to the product. Waste is generated especially when people, equipment, supplies, tools, documents, or materials are moved or transported unnecessarily from one location to another; examples include

transporting wrong parts, sending materials at the wrong time or to the wrong location, transporting defective parts, and sending documents that should not be sent at all.

7. Unnecessary work performer motion:
Unnecessary motions of workers are caused by poor workstation layout, disorganised workplace and storage locations, and non-standardised processes and material flows. All wasteful work performer motions cost the organisation time (money) and can cause stress on employees which could lead to injuries or employees making mistakes.

5S/5C:

A welcome find in some manufacturing companies in China are 5S/5C practices (or 6S for those organisations that choose to add "Safety"). This set of practices is regarded as a practical foundation for stabilising, maintaining and improving work-areas, supporting quality management processes, and providing the correct environment for Lean Manufacturing. In the West, 5S/5C is usually the first practical "Lean" foundation method which organisations implement.

Upon seeing the 5S/5C/6S practices in organisations in China, the impact to the visitor is pleasing because workstations appear to be in a state of control; they are visually organised and clean. A deeper look reveals that, generally, the 5S/5C/6S practices are not as inclusive of quality practices as they are in the West and important elements appear to be missing notably pertaining to, for example:

- document control
- identification of abnormal conditions, and broken and safety critical items
- identification of quality affecting conditions
- work instructions, key point charts, visual aids, visual standards
- quality status identification of components
- on-the-job skills and work-practice training
- Personal Protection Equipment (PPE)

Table 8-1 shows the aim of each 5S/5C practice applied to production and office areas and typical "How" actions as found in Western organisations. Contained within these practices are quality practices that have been around long before the rise in popularity of 5S/5C.

Table 8-1: Typical 5S/5C practices and actions

1. Sort or Clear-out Aim: To have in the work-area only what is required for the work. How: Separate the essential items from the non-essential items.
2. Set in Order or Configure Aim: To have a place for everything and for everything to be in its place. Make it easy to find things and to identify incorrect and abnormal conditions. How: Actions include locating components, tools, gauges and equipment in identified and ergonomically efficient, accessible and safe positions in the work-area, and setting in place well-thought-out ways of doing things; this involves: - manufacturing methods (procedures, work instructions, key point charts), - production tooling and quality inspection instruments, - shadow boards and cabinets for the necessary tools and measurement gauges, and storage for cleaning materials, - drawing, work instruction, and procedure control (removal of old and obsolete documents, and also old and obsolete notes from notice boards), - consistent content of documentation accompanying the job, - clerical processes for job delivery to the workstation and for progress tracking, - area zoning and floor marks (by means of painting/colour coding) for material storage, personnel safety and access control
3. Shine or Clean and Check Aim: To assess the condition of area and/or machine and keep it in good order. How: Manually clean. Identify any items (e.g., with a "Red Tag"); these include quality critical, safety critical, broken and homeless items. Identify concerns as well as any abnormal conditions. Identify status of material and ensure non-conforming material is segregated from conforming material.

4. Standardise or Conformity

Aim: To ensure that the 5S/5C practice is maintained and improved.

How: Make it easy to maintain the 5S/5C condition with:

- daily checklist
- visual management for control and standards
- clear assignment of responsibilities

5. Sustain or Custom and Practice

Aim: Employees understand the benefits and make 5S/5C "the way of life".

How: Provide management support and encouragement to secure on-going commitment. Practical actions are those such as:

- on-the-job skills and work-practice training
- quality education
- PPE and personal health and safety training
- making job descriptions
- making work instructions
- performing audits
- recognizing and rewarding desired behaviour
- daily management attention and interest

Note: It is essential to address 4 and 5; if neglected, the area will eventually revert to a non-5S/5C state.

In successful implementations of 5S/5C, the journey has begun with absolute and full management support. To demonstrate what can be achieved, the initial launch has often been in a good example work-area. This tactic tends to open the way in gaining people's trust and support. Thereafter the implementation has been gradually rolled-out to other work-areas in the organisation.

Organisations have particularly reaped great benefits when implementation has been accompanied with the teaching of an understanding of the principles of "Lean Thinking" and the "7 Wastes"; all people involved in the 5S/5C implementation have been given an explanation of company objectives, the rules of 5S/5C practices, the principles of "Lean Thinking" and "7 Wastes".

Notably, most of the necessary practical knowledge has been imparted in the work-area through a facilitator working with the people in the work-area. This follows because 5S/5C is work-area specific and must be taken up, practised and owned by the people that work within the work-area.

The benefits that 5S/5C implementation brings to an organisation can be significant; these stem from the large number of individual benefits provided by the various 5S/5C practices and are summarized as follows:

- Practice 1 – Sort or Clear-out:
  The removal of all unnecessary items (e.g., documents, tools, components) from the work-area results in less clutter and promotes orderly thinking; needed items such like tools, jigs and measurement gauges are quicker and easier to find. In addition, removal of unwanted obstacles increases motion efficiency and safety.

  Furthermore, people working in the area are sensitive to non-moving components visible for example in the stacking-up of work awaiting quality disposition or further processing. This tends to spur the taking of action.

- Practice 2 – Set in order or Configure:
  With components, equipment, tools, jigs, and measurement gauges located in clearly identified and ergonomically efficient, accessible and safe positions in the work-area, motion efficiency and safety is increased, quicker changeovers are made possible, the opportunities for handling damage is reduced, and time is not wasted searching for things.

  Inadvertent incorrect processing of components is prevented when an area is designated and clearly identified to temporarily hold components awaiting inspection or components found non-conforming and awaiting quality disposition.

  The setting in place of consistent and well-thought out ways of doing things minimizes the opportunity for errors and of quality-related risks from occurring, and schools people in good habits.

- Practice 3 – Shine or Clean and Check:
  The practice of cleaning or "laying hands on" the work-area, and machines, tools, jigs and measurement gauges in the work-area, and checking everything, brings the benefits of early identification of any

quality critical, safety critical, broken and homeless items, as well as any concerns and abnormal conditions. This leads to actions being taken to prevent problems from escalating and of serious breakdowns and other delays from occurring.

The practice of making sure that the status of material is identified, or that the material is placed in a clearly identified location, prevents the inadvertent use of non-conforming material.

- Practice 4 – Standardise or Conformity:
  With people in the work-area maintaining and improving their practices to achieve the most efficient way of working, a less wasteful and more efficient working environment is developed.

  Prevention of problems and prevention of waste is facilitated through people in the work-area routinely ensuring that the correct methods, standards, and checks are employed, and the right tools are available.

- Practice 5 – Sustain or Custom and Practice:
  This ongoing practice brings the benefits of ensuring that gains obtained are sustained and further enhanced.

  The attention to job details, clear work instructions, and training and education in necessary work-skills, safety and quality, enhances the effectiveness of people in the work-area.

  The ongoing support, encouragement and recognition from management bring the benefits of sustaining motivation and the positive work environment.

5S/5C provides a solid foundation for achieving operational excellence.

The gains made through successful 5S/5C implementation cost the organisation a minimal amount. Overall there are improvements in efficiency, quality, personal safety, and employee morale in the workplace. The tidy and organised layout of work-areas, resulting from 5S/5C implementation, also boosts the "quality" image of the organisation.

Questions for Chapter 8

8-1: Discuss how processes are most effectively arranged to satisfy customers' needs and create value.

8-2: Summarize key activities as found in better organisations that enable, support and continuously improve operational performance.

8-3: Discuss the advantages and disadvantages of piece-work schemes.

8-4: PDCA is an improvement method and a way of thinking. Explain the phases of PDCA.

8-5: A manufacturing organisation has refurbished an automated welding machine and wants to know whether refurbishment has improved OEE.

Before refurbishment, the machine had the potential to make 15 butt-welds per minute. The machine was planned to operate 18 hours out of every 24 hours, however there were frequent stoppages due to machine sequencing problems. The result of these stoppages meant that the welding machine was actually operating for 12 hours 10 minutes on average out of 24 hours. During this operating time, 9860 welds were made and 9120 welds passed quality control tests.

After refurbishment the butt-weld cycle time increased resulting in the potential butt-weld rate being reduced to 12 butt-welds per minute, but stoppages decreased dramatically and an actual operating time of 15 hours 20 minutes out of a planned 18 hours was achieved. During this time, 10190 welds were made and 10156 welds passed quality control tests.

Use the given information and equations in the table below to calculate the Availability %, Performance %, Quality % and OEE % before and after the changes. Discuss the results and suggest what could be attempted to further improve OEE.

Equations		Before	After
Availability % =	$\dfrac{\text{actual operating time}}{\text{planned operating time}} \times 100$		
Performance % =	$\dfrac{\text{cycle time x total units started}}{\text{actual operating time}} \times 100$		
Quality % =	$\dfrac{\text{good units produced}}{\text{total units started}} \times 100$		
OEE % = Availability % × Performance % × Quality %			

8-6: Discuss the following causes of the cost of poor quality: (a) Non-conforming product found during final inspection or testing; (b) Past due date receivables; (c) Expediting.

8-7: Briefly explain the understanding that is helpful in promoting "Lean Thinking" in a manufacturing organisation.

8-8: Identify quality-related benefits an organisation can secure from implementing the 5S/5C practices.

9. SUPPLIER QUALITY ASSURANCE

The average manufacturing organisation spends typically over 50% of its income revenues on purchased inputs. Expediting costs, discussed in Chapter 6.2 and 8.3, can seriously inflate the cost of these purchased inputs. It is clearly a strategic necessity for manufacturing organisations to aim at having suppliers that can deliver quality conforming product on-time to an agreed schedule, and at a consistently competitive cost, and a supplier quality assurance programme has the purpose of providing confidence that the products or services delivered will satisfy the purchaser's quality, time and cost requirements.

It is encouraging to find leading industrial manufacturing organisations in China with comprehensive sub-contractor and bought-in product management directive documents that contain the principles to follow for supplier evaluation and selection, conditions of supply, delivery planning, quality control, pricing, payment and discipline. These organisations maintain an approved suppliers list; they monitor defective product and delivery shortcomings, and frequently engage with suppliers regarding these issues. Such industry leaders operate in a challenging environment in that they are often left with the difficult choice of having to work with developing suppliers – suppliers that have the basic capability to produce what the purchaser requires, but lack the expertise to do the work to the specific and often exacting requirements of the purchaser.

This chapter includes the approaches and practices that have been seen to be successfully employed in supplier quality management in both developing and developed economies, and includes elements of supplier quality management that are considered to be of benefit to industrial manufacturing organisations in China; these are explained under the following headings:

1. Supplier selection
2. Arranging what needs to be controlled
3. Ensuring the purchaser gets what is specified and agreed
4. Reviewing and rating supplier performance
5. Supplier development

9.1 Supplier selection

The supplier selection process begins with the purchaser having a specific requirement that includes technical specification, quantity, delivery, and at the initial stage, approximate cost. For some purchases, the supplier selection process is straightforward; for instance, purchases that are sufficiently defined to allow selection from a catalogue such as personal protective equipment (PPE) and office supplies. However, for the purchase of complex manufactured goods, detailed supplier appraisals or assessments are usually generally needed to assess the potential supplier's or sub-contractor's capability to do the required work, and their abilities to control quality, cost and delivery schedule. The purchaser also needs to know the supplier's limitations that could present possible risks in meeting requirements.

The discovery of these facts is through a formal "Supplier Quality Assessment", also called a "Vendor Evaluation". This is undertaken by a team, usually from procurement, technical and quality, that are knowledgeable with respect to the technical standards, specifications and processes that the supplier should have in place to provide a product or service that meets requirements. The team must also have the experience to be able to determine the suitability of the supplier's quality and management processes, and the organisation that the supplier should have in place to <u>consistently</u> meet technical, cost and delivery requirements.

Prior to visiting the potential supplier's organisation to conduct a "Supplier Quality Assessment", a "Supplier Questionnaire" is sent, customarily with an assurance of confidentiality with respect to the answers given. Questions pertain to the particular supply requirement, and concern production equipment and facilities, production capacity, quality organisation, process control, inspection and test systems, financial position, and possibly environmental management.

The answers should provide information to determine the viability of the supplier to do the work, they should draw attention to specific strengths and possible limitations of the supplier, and also identify whether a similar assessment by another party has been done recently. The answers are taken into account in the decision to visit the supplier.

Visits to suppliers are necessary when the potential supply is for complex goods, high value/high risk items, a major contract, or when special controls are needed. The answers to the "Supplier Questionnaire" are used by the team to prepare themselves for the visit; this includes the drafting of a checklist (an aide-memoir) to ensure that important questions are not overlooked during the visit.

Sometimes the purchasing organisation may employ independent consultants to perform supplier appraisals.

Findings and discussions with the supplier's personnel during the "Supplier Quality Assessment" visit will typically reveal or provide an indication of the following:
- Attitudes of the supplier's employees towards their work and the control of its quality.
- Capability and adequacy of production equipment to meet the purchaser's requirements.
- Care and maintenance of production equipment (relates to quality consciousness).
- Technical knowledge of supervisory staff and their ability to control operations.
- Suitability and adequacy of inspection and test methods employed during production processes.
- Adequacy of the supplier's quality management processes and practices for controlling, e.g., technical documents, non-conforming material, purchaser's supplied tooling.
- Orderliness and cleanliness of the supplier's production premises (gives an indication of management's planning and control).
- Attitude towards safety and disaster avoidance (relates to continuity of supply).
- Expertise of personnel regarding providing solutions to process problems.

- Commitment of the supplier's management to quality.

The results of the "Supplier Quality Assessment" will indicate what strengths and weaknesses the candidate supplier has and whether weaknesses could cause problems for the purchaser. The results will also help to identify risks so that they can be effectively managed.

Industrial manufacturing organisations typically classify suppliers in terms of ability and sourcing conditions. This type of classification regime was found in leading manufacturing organisations in China where it was not uncommon to find, on their "List of Suppliers" a three-level classification, usually as follows:

A: Supplier has the required engineering ability to manufacture certain products to specified acceptance criteria, and a stable co-operation relationship is established.

Classification A suppliers can be used for routine sourcing.

B: Supplier has the required equipment to manufacture certain products and the potential to develop into an A classified supplier, but needs assistance to strengthen production process control, quality supervision and/or delivery performance.

Classification B suppliers qualify for use under guidance and advice.

C: Temporary supplier. This classification of supplier may be used only when severe bottlenecks occur.

The supplied product must be strictly inspected by the purchaser before despatch of product from the supplier's premises.

9.2 Arranging what needs to be controlled

Specified purchase requirements are conveyed via a "Contract to Supply", or a "Purchase Order" that carries details such as product description, required quantity, delivery time/schedule, delivery address, payment conditions, technical requirements (referencing, e.g., a drawing, a material specification, a standard for inspection, and testing acceptance criteria), as well as conditions deemed necessary from analysis of the "Supplier Quality Assessment" findings.

The findings contained in the assessment report are analysed and used not only to determine supplier suitability, but also to identify contractual conditions of supply pertaining to quality control and assurance that may include:

- Stipulation that certain work such as welding and Non-Destructive Testing must be performed by certificated personnel.
- Notification that the supplier will receive frequent "Surveillance Visits" from the purchaser's Quality Assurance Representative (QAR); these are quality control-orientated visits to the supplier's factory.
- Instruction to the supplier that product must be released from his premises only by the purchaser's QAR.
- Request for all non-conforming products to be communicated to the purchaser's QAR.
- Instruction that evidence of specific process controls are supplied with each delivery, e.g., hardness tested samples after heat treatment, with proof of Brinell test indentations.

These conditions of supply are as a rule written into the "Contract to Supply".

For goods requiring special controls, the purchaser may require the supplier to follow a Quality Plan; see table 9-1 for a summary of requirements of a quality plan for bought-in product. The purchaser could request the supplier to draw up the quality plan, or the purchaser may draw it up. For the supplier to draft the quality plan, the supplier will need to find out or be advised what controls are needed to meet the purchaser's quality requirements. Quality plans are mostly always discussed and agreed between both parties.

In industrial manufacturing organisations in China were seen documents named as quality plans given by the purchaser to the supplier. These typically contained broad requirements, for instance, "make to GB/T standard xyz" and did not contain control requirements such as those identified in table 9-1.

Table 9-1: Quality plan for bought-in product - summary of requirements

---

- The Quality Plan contains actions for the supplier to follow during all activities within the scope of the supply agreement. Should the supply agreement include development, manufacture, delivery and installation, then the Quality Plan will apply to all of these activities.
- Actions could call for the following:
  - application of specific controls,
  - application of specific process control work instructions,
  - use of reference samples, visual aids and workmanship standards,
  - adherence to special procedures and purchaser supplied specifications,
  - adherence to requirements stipulated in national or international standards,
  - performing of specific inspections and tests at certain stages.
- Hold and witness points could be identified with conditions as follows:
  - At "hold" points, the supplier cannot proceed without the approval of the purchaser, or his representative.
  - At "witness" points, the supplier is required to arrange for the purchaser, or his representative, to observe a certain activity.
- The Quality Plan usually stipulates requirements for "record of proof" that the supplier must retain as evidence that certain special process controls, and inspections and tests have been carried out.

The better of the quality verification documents seen in China for bought-in products was an "Inspection and Commission Procedure" for the sub-contacted build of a complex industrial product. This document identified what must be inspected and tested by the sub-contractor, and what proof of inspection and test was required – certainly a sensible and valuable document.

9.3 Ensuring the purchaser gets what is specified and agreed

When the implementation of the sensible "Inspection and Commission Procedure" mentioned in Chapter 9.2 was verified at a later audit, the stipulated inspection result record as called for in the procedure was found to be blank – the requirements of the "Inspection and Commission Procedure" had never been followed through.

Unfortunately, this was not an isolated incident. It should be needless to say that it is imperative for the purchaser to follow through the implementation of all contractual requirements and conditions of acceptance to ensure that he is getting what was specified and agreed.

All conditions of acceptance and contractual requirements must be readily accessible by the purchaser's QAR or the goods receiving personnel given the task of acceptance. Requirements and conditions of acceptance are usually traced (and "communicated") on computer through the purchase order number. The routine practice of checking that all purchases are verified against the specified and agreed contract requirements can occur during visits to the supplier's or sub-contractor's premises, or upon receipt at the purchaser's stores.

The purchaser may need to perform "Surveillance Visits"; during the visit, the purchaser's QAR may view processes being carried out, perform random inspections and measurements, and make observations with regard to potential risks. For example, the QAR may observe a lack of evidence of the control of a certain production process, or see out-of-calibration equipment, or observe an incorrect (or absent) operator action, or find a material traceability issue.

The purchaser's QAR may be guided by a contractually invoked quality plan that stipulates "witness" and "hold" points. At "witness" points, the purchaser's QAR is required to observe a certain activity, e.g., a welding process, a heat treatment process, or a component test. At "hold" points, the supplier cannot proceed without the approval of the purchaser. A "hold" point could be defined, for instance,

- immediately prior to assembly of certain components because the purchaser wants to perform his own inspection or test and cannot do this after assembly,
- or prior to painting to check the condition of the unpainted surface for evidence, e.g., of hairline cracks,
- or before shipment to site from the supplier's premises.

A good practice is to write down observations in a "Quality Visit Report" at the time of the visit, and to give the supplier a copy of this report prior to leaving the supplier's premises. The "Quality Visit Report" has proven to carry many advantages over verbal communication. The observation, when written down, provides clearer expression, and the report formally communicates the problem with the same meaning to all concerned, and also enables specific follow-up to be performed, if not by the QAR, then by a work colleague.

If the purchaser's QAR has performed successful product release inspections and tests at the supplier's premises, and there is proof of this in the delivery, the extent of examination upon receipt at the purchaser's stores is limited to the possible effects of packaging and transportation. In cases where QAR release at the supplier's premises has not been performed, and where the supplier has not been qualified to "Ship-to-Stock" (discussed in Chapter 9.4), the products undergo a receiving inspection process upon receipt at the purchaser's premises.

In order to perform an adequate verification, the person performing receiving inspection has, for each product, drawings or specifications marked-up with critical characteristics, and sampling tables or sampling requirements from applicable standards. The inspector also refers to the "Supplier Quality History Records" for the performance of previously delivered products from the supplier. If these records indicate many non-conformances or other repeat problems, examination could be increased (referred to as tightened inspection). On the other hand, if the supplier demonstrates an exceptionally good record of delivering conforming product or material, the verification could be relaxed (referred to as reduced inspection).

When a non-conformance is found during receiving inspection, a non-conformance report is generated and appropriate personnel are requested to decide the disposition. The non-conformance is formally communicated, in writing, to the supplier – some organisations use a pro-forma "Supplier Corrective Action Request" notice. All rejection details are recorded in a "Supplier Rejection Register".

Certain non-conformances could lead to the QAR visiting the supplier's premises to examine the situation and discuss appropriate controls that the supplier needs to implement to prevent recurrence. Repeated and serious non-conformances could prompt the purchaser to review the "Contract to Supply".

9.4 Reviewing and rating supplier performance

Good sources of information concerning suppliers' performance that are being continually updated are the "Supplier Quality History Record", the "Supplier Rejection Register", "Quality Visit Reports" and delivery records. The purchaser uses these in the periodic review and "rating" of his suppliers. The motivation of the purchaser for establishing a supplier rating system is part of the effort to ensure that its suppliers develop the desired characteristics and abilities to provide products and services that meet all of the purchaser's requirements.

The supplier performance review enables the purchaser to identify good and poor supplier performance, and to take appropriate action. Consistently good performance could mean that the purchaser upgrades the supplier rating and relaxes surveillance at the supplier's premises, or reduces the amount of receiving inspection. Some organisations in the West allow suppliers with very high supplier ratings to "Ship-to-Stock"; this means that qualified products or components can be shipped directly from the supplier to the purchaser and are ready to be fed to the purchaser's production line upon delivery. This supports Lean Manufacturing for both purchaser and supplier.

Poor performance could mean that the purchaser downgrades the rating of the supplier and applies more stringent controls. Very poor performance may lead to the purchaser disapproving the supplier and therefore no longer placing purchase orders with him.

Typical supplier performance review criteria used in a supplier rating system are as follows:
- Overall quality conformance and appearance of the supplied product or service.
- Frequency of rejection of deliveries upon receipt.

- Supplier's products or services causing the purchaser's customers to complain.
- Maintenance of the price of product or service at a competitive level.
- Adequacy of required paperwork accompanying deliveries (e.g., inspection and test record certificates, and material certificates).
- Supplier's history of meeting delivery schedules.
- Stability of the supplier's financial situation.

Each of these criteria is assigned a weighted maximum score. Demerits are given for poor performance, and an earned score is calculated. The "Supplier Rating", also referred to as a "Vendor Rating", is determined from the sum of the earned scores for these criteria.

The "Supplier Rating" is entered onto the "List of Suppliers" which contains all suppliers of products and services, and the purchaser refers to this list of suppliers for all purchases. Typically, the information shown in table 9-2 is held for each supplier on the "List of Suppliers".

Table 9-2: Typical information held in a "List of Suppliers"

1.	Reference Code (for the supplier)	
2.	Name of Supplier	
3.	Classification of Supplier	(A) Used for routine sourcing; (B) Used under advice; (C) Used only under special circumstances, and under advice
4.	Supply Category	Material; Product; Packaging; Transport; Maintenance; Tooling; Services
5.	Status	Approved and supply contract in place; Approved; Disapproved; No alternative
6.	Supplier Rating	
7.	Comments These may be, for example: details of specific products for which the supplier is approved, whether a "Quality Plan" is in place, or whether "Surveillance Visits" need to be performed.	

The frequency of supplier performance reviews is usually no less than one on each supplier every year, but this may be increased by such things as criticality of the product, or disturbing trends seen in the "Supplier Quality History Record" and the "Supplier Rejection Register".

9.5 Supplier development

In order to improve supplier rating, purchasers often invite selected suppliers to participate in supplier development programmes. In these programmes, the purchasing organisation assists selected supplier organisations to improve their performance. Of course this is done primarily for the benefit of the purchasing organisation but there are usually significant benefits to the supplier organisation.

Supplier development programmes are used in both developed and developing economies and are particularly effective in situations where the purchaser is driven by circumstance or by lack of choice to source product from small and medium sized suppliers that have the basic capability, but lack the expertise.

The involvement of the purchaser in these programmes has been seen to range from a little to a lot. Whatever the degree of involvement may be, the commitment of both the purchaser's and supplier's management has been the main success-influencing factor.

The commitment of the supplier's management is usually secured through a business agreement in which the purchaser is able to convince the supplier's management of the benefit and value of the supplier committing his resources to the required improvements. The purchaser must ensure that the supplier does not see the agreement with the purchaser as coming at too high a price or at the expense of other business needs, because this is sure to limit the success of the purchaser's efforts.

The main assistance given by the purchaser to a supplier is through manpower. This is not through providing additional manpower to do the contracted work, but through selected expert manpower that brings to the supplier knowledge and guidance in practices, processes and techniques commonly on an as-required part-time basis. The amount of manpower and the nature of skills will depend on the situation – occasionally one or more specially skilled personnel are sent to work in a guiding or teaching capacity at the supplier's premises for three to six months, sometimes the purchaser's employees may only be required to provide guidance at the supplier's premises for half a day a week.

The assistance can be of a general or specific nature, and will depend on factors such as the development status of the supplier, potential risks faced by the purchaser, and weaknesses in the supplier's control systems. General assistance could be provided by way of guidance in, for instance, workmanship standards, practices to eliminate the 7 Wastes, the use of Basic Quality Tools, visual displays for important measurement information, visual controls (e.g., to guide actions and to identify components), and 5S/5C work-area organisation. Specific assistance could be provided, for instance, in guiding the application of applicable technical process specifications, or in explaining the practical implementation of requirements contained in national and international standards, and in the guidance and training in the application of contract-specific processes.

It is extremely unusual to find a purchaser providing direct financial assistance to a supplier. The purchaser may supply costly tooling, e.g., injection moulding or casting dies, but this is not considered part of supplier development.

In large manufacturing organisations it is quite common to find, usually within the Quality Department, a dedicated supplier development unit. With such a unit, the purchaser can provide a focused approach to improving the quality and delivery performance of suppliers.

A well-organised supplier development unit would have prepared or acquired the necessary effective "tools" for the job. These could include the following:

- Guidelines or templates for quality plans
- Examples of visual displays for measurement information and also a supplier development scoreboard
- Examples of visual controls (refer to Chapter 3.3)
- Workmanship standards
- Technical process specifications
- National and international standards (GB/T, DIN, EN, BS, etc.)
- Guidelines for 5S/5C work-area organisation (refer to Chapter 8.4)
- Practices to eliminate the 7 Wastes (refer to Chapter 8.4)
- Guidelines for the use of Basic Quality Tools

A supplier development scoreboard, located at the supplier's premise, can provide encouragement when metrics are chosen relating to the purchaser's requirements and that are appropriately meaningful to the supplier.

Scorecards reflect the supplier's performance in areas that really matter. Opportunities for improvement are identified and the purchaser then works with the supplier to improve these areas.

As the supplier's performance improvement progresses – reflected in improved quality conformance and improved delivery reliability and punctuality of incoming goods – so does his "Supplier Rating". Many organisations have awards associated with supplier ratings that indicate the level of supplier performance excellence. Certain leading organisations have "Preferred Supplier Programmes" with award levels designated by way of numbers of stars or named, for instance, Silver, Gold and Platinum. Each award level exhibits a more stringent level of performance excellence. A Five Star and a Platinum level would be awarded to the supplier who has shown consistency in attaining superior performance over a number of areas that include achievement of 100% quality conformance and 100% on-time delivery.

Supplier Development requires purchaser and supplier to commit resources to the development work, and often there is a need to share information which could sometimes be sensitive. The programme is challenging for both parties and requires management support from supplier and purchaser, but the benefits to both parties can be substantial.

Questions for Chapter 9

9-1: A potential supplier is identified for the possible supply of complex, high value items for a major contract. What should the purchaser's supplier selection team be on the lookout for when they assess the suitability of the potential supplier?

9-2: Identify the various quality control and quality assurance conditions the purchaser can stipulate in the "Contract to Supply".

9-3: Outline possible reasons for the purchaser to have a Supplier Development Programme.

9-4: With regard to the purchaser wanting to develop a supplier, describe the typical assistance that the purchaser can give to the supplier.

PART 5: VARIABILITY, CAPABILITY & QUALITY TOOLS

Subjects, practices, tools and techniques contained in Part 5

Chapter 10: Process and product variability and capability
- ♦ Process variation – common and special causes
- ♦ Machine or process variability and capability
- ♦ Process Capability Index
- ♦ Product capability
- ♦ Design safety margin
- ♦ Probability of failure
- ♦ Decrease of product strength over time

Chapter 11: Basic Quality Tools
- ♦ Basic Quality Tools: Flow Chart, Check Sheet or Defect Tally Sheet, Histogram, Pareto Chart, Cause and Effect Diagram, Scatter Diagram or Run Chart, Control Chart
- ♦ Control Chart: Principle of operation, $\overline{X}$ and R Control Chart construction and interpretation
- ♦ Estimating a tolerance that can be maintained by an in-control process

10. PROCESS AND PRODUCT VARIABILITY & CAPABILITY

Too often in manufacturing organisations in China issues were encountered concerning process and product variability and capability. These would arise, for example, when difficulty was encountered maintaining tight tolerances specified in technical documents or when a production machine indicated an incapability of maintaining a specified accuracy; declared rejects were constantly high and "accept on concession" permits in the factory were continuously being granted. Product capability would particularly become a concern when product or equipment was specified without due regard and understanding of its application or the environment in which it should operate; equipment breakdown resulted, and the equipment manufacturer had to replace the failed items in the field at their cost.

These issues stem from a number of reasons, foremost of which are the following:

- The designer or specifier had not considered the capability of the production machine or process with regards to doing the job.
- The inexperience of the designer caused him to be cautious and specify unnecessarily tight tolerances.
- Distrust in work integrity of production operators caused the designer to specify overly tight tolerances on drawings and technical specifications.
- The equipment specifier did not appreciate the full nature of the stresses that would be encountered by the product or equipment during its operation in the field.

These reasons suggest a lack of an adequately deep appreciation of variability and capability. This chapter therefore presents information on the following:

1. The nature and causes of process variability
2. Getting to know process variability and capability
3. Process Capability Index
4. Capability of the product in its intended application

10.1 The nature and causes of process variability

The four pictures in figure 10-1 represent various types of process variabilities indicated as groupings of points on targets. Each picture includes the distribution curve representing the grouping on the target. The Upper Specification Limit (USL) and Lower Specification Limit (LSL) define the allowable deviation from target-centre. The process average ($\bar{x}$) is indicated on the distribution curve.

In picture (1) the aim is set on target-centre and the grouping of points is central to the target-centre. The points are spread all over the target, and many points are not within the specification limits. These out-of-specification points are rejects, defects or failures. The frequency distribution curve displays this large variation on either side of the average ($\bar{x}$). The process does not have the capability of performing 100% within specification limits.

In picture (2) the aim is set to the left of target-centre. The grouping of points is tighter than those in picture (1), but to the left of the target-centre. A number of points are out-of-specification and are therefore rejects, defects or failures. The frequency distribution curve displays a variation smaller than that in picture (1) on either side of the process average ($\bar{x}$). If the aim were to be reset to target-centre, the process would be capable of performing 100% within specification limits.

In picture (3) the aim is set on target-centre and the grouping of points is tight and central to target-centre. There are no points beyond the LSL and USL and therefore there are no rejects, defects or failures. The frequency distribution curve displays a small variation on either side of the average ($\bar{x}$). The process is capable of performing within specification limits.

In picture (4) a similar grouping to that of (3) is shown, but a few points are out-of-specification, beyond the LSL. The process appears to be potentially capable of producing 100% within specification limits, but something is causing a few points to not fall within specification.

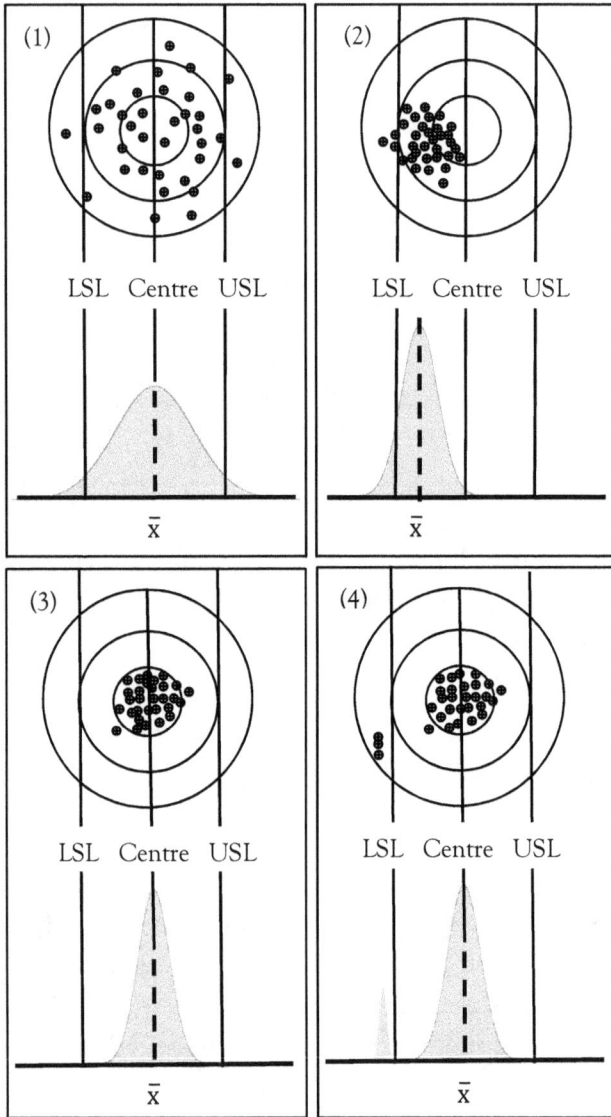

Figure 10-1: Illustration of process variability and capability

The frequency distribution curves in pictures (1), (2) and (3) of figure 10-1, illustrate a numerically measureable process operating in a stable and predictable state. The natural variation arises from causes inherent to the process, and the process results are distributed in a bell-shaped or

"Normal" frequency distribution curve. The "Normal" frequency distribution arises from "common" causes of variation, e.g., an oven's thermostat regulating oven temperature.

The frequency distribution curve in picture (4) of figure 10-1 illustrates an effect of a "special" cause of variation; the outlier points could have been caused by, e.g., opening the oven door during heat treatment causing the oven temperature to suddenly drop.

Some examples of "common" and "special" causes of variation are given in table 10-1. Most of the issues that can be classified as "special" causes of variation are largely workforce controllable, whereas most "common" causes of process variability are generally due to system or management-controllable issues. Efforts towards process improvement must be focused firstly on removing "special" causes of variation.

Table 10-1: Some "common" and "special" causes of variation

"common" causes:	"special" causes:
• an oven's thermostat regulating oven temperature • an operator making an occasional error • variation of raw material properties • excessive machine wear and tear • poor maintenance of machines • poor working conditions • variation in delivery time during normal traffic conditions	• opening the oven door during heat treatment causing the oven temperature to drop • an operator absent or not paying attention • incorrect raw material • tool breakage or machine malfunctioning • machine needing adjustment • a new untrained operator making many errors • no working procedure where needed • abuse of equipment

10.2 Getting to know process variability and capability

With respect to being able to perform within specification limits, it can be seen by way of figure 10-1 that we need to have a measure of process variability and capability. Knowing this enables us to direct an appropriate level of quality control effort to ensure that out-of-specification situations are avoided or detected and corrected in a timely manner. Having a measure of process variability and capability also opens the door to quality improvement. We therefore need to spend a little time obtaining a measure of the performance of our machine or process. An appreciation of the performance can be gleaned over a few work shifts by measuring and plotting process variables on a Run Chart. (See 11-6 for explanation of Run Charts.)

The measured process variable could be the machined diameter on a lathe, the amount of weld-upset in a weld, the temperature of components undergoing heat treatment, the coil resistance of coils from a coil-winding machine, etc. The measured variable is plotted on the y-axis, and successive pieces (or time) plotted on the x-axis. A graphical pattern of variation will emerge that can be compared with six possibilities illustrated in figure 10-2; this will help in deciding on the appropriate level of quality control effort.

Chart 1 in figure 10-2 illustrates a machine that produces successive pieces with small variation. This variation is similar to that depicted in picture (3) of figure 10-1. The specification limits, shown in solid lines, are set at 16.0 and 16.4 mm. The natural variation inherent to the machine, and due to "common causes", is such that it is very capable of producing within the specification limits, and will only need a minimum amount of checking to make sure that this performance is maintained. However, if the specification limits were tightened to the dotted lines drawn at 16.1 and 16.3 mm, the machine would not be capable of producing 100% within these tighter limits.

Chart 2 illustrates a machine that produces successive pieces with little variation then suddenly something goes wrong and the setting changes to a new level. The average value changes but the process variability remains the same. This sudden change is the result of a "special" cause such as the setting being "bumped" to a new level after a tool jam, or, following a change in heat treatment operator and the new operator not adhering properly to the process procedure, or after the start of a batch of material with a big difference in properties, or the process suddenly picking up an impurity, etc.

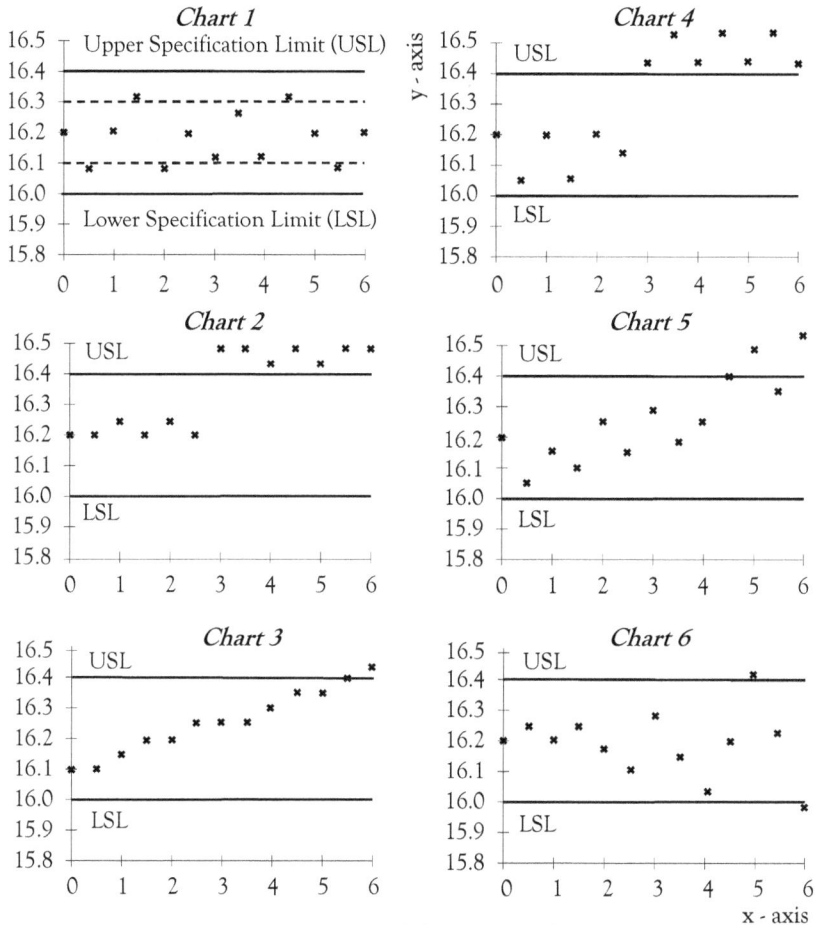

Figure 10-2: Types of variability in a machining process

Chart 3 illustrates a machine that steadily drifts upward. The average value steadily drifts but the process variability remains the same. This issue is mentioned in Chapter 3.2 where the welding electrodes deformed under successive pressure cycles and wore down causing the amount of weld-upset to gradually decrease in successive links. There can be many other "special" causes such as tool wear, or chemical solutions becoming spent, or progressive contamination of a chemical bath through impurities being introduced by successive work-pieces.

Sometimes drift is unavoidable and, having identified this pattern, the moment that the process goes out of a specification limit can be predicted using a chart.

For machines or processes having performance that is typified by the variation patterns illustrated in Charts 1, 2 and 3, only simple quality control is required because the variation is small and attention needs only to focus on the average. The last piece that is checked will give a good indication of quality status between checks. Only work made since the last check will need to be sorted if the last piece is out-of-limits.

Machines or processes that perform as illustrated in Charts 4, 5, and 6, and those that have other types of patterns where variation is appreciable, are best controlled using a control chart such like the Averages and Range Chart discussed in Chapter 11.7.

Charts 4 and 5 illustrate a type of performance similar to that depicted in Charts 2 and 3 respectively, but the process has a lot more variability. Elimination of the "special" causes leading to the sudden change in setting (Charts 2 and 4) and the steady drift (Charts 3 and 5) should be prioritized.

Chart 6 illustrates a machine or process where the inherent variability is initially small but gradually increases over time to a stage where the process is no longer capable of producing within specification limits. The cause of increasing variability over a long time could be, for instance, due to increasing play in machine bearings – on machine work the variation often indicates the general overall condition of the machine.

On expensive and reputable brand-name equipment the inclination observed in industrial manufacturing organisations in China was to trust this equipment to produce perfect quality once set up correctly.

No matter how good the machine or process variability presents itself and how capable the machine or process appears to be with regards to working within specification limits, it must not be assumed that once set up, the machine or process will continue to produce to the same average and standard deviation indefinitely. The reality is that tools wear, swarf can get in between the machine stock and work-piece, material variation can occur, etc., therefore some degree of quality control is always required.

10.3 Process Capability Index

The designer or specifier should have a good idea of the capability of the production machine or process with regards to doing the job, i.e., he should have a good idea of the capability of the machine or process to perform within his drawing or specification limits. Information on machine or process performance could be available to him through the history records of similar jobs, or he or the quality engineer may need to spend a little time obtaining a measure of the performance of the machine or process.

A measure of the performance can be achieved by way of a process capability study. This requires that the process performance measurements of a stable and in-control process are compared with specification limits to determine the capability of the process to perform within the specification limits.

The validity of process capability measurement is dependent upon three conditions:

1. The process must be operating in a stable and predictable state and there must be no "special" causes of variation indicated by way of, for example, the outliers shown in (4) of figure 10-1 and the issues illustrated in Charts 2, 3, 4, 5 and 6 of figure 10-2.
2. The process results should be distributed in a bell-shaped or "Normal" distribution within three standard deviations (σ) on either side of its process average ($\bar{x}$).
3. The sample measurements should be randomly selected and numerous enough to accurately represent the process.

The stability and predictability of process variation is usually confirmed using a quality control chart such like the Averages and Range Chart over a sufficiently long period of time to build confidence that "special" causes of variation have been removed, and to ensure that time-of-day/week/month or time-of-year effects are not biasing measurements.

The ratio of the process performance measurements to the specification limits is referred to as a Process Capability Index (Cp and Cpk).

A Process Capability Index expresses how well the process is performing compared to the specification limits. This is explained referencing figure 10-3 below, where:

- Cp is the ratio of the specification width to the process variation width: $Cp = a/b$

  Since Cp does not take account of how well the process distribution is centred within its limits, Cpk is used to provide a more accurate Process Capability Index ("k" stands for "centralizing factor"); the smaller of c or d is divided by half the process spread: $Cpk = c / \frac{1}{2}b$

  (When the process distribution is centred, namely, $c = d = a/2$, therefore $Cpk = Cp$.)

Lower Specification Limit (LSL)   Upper Specification Limit (USL)

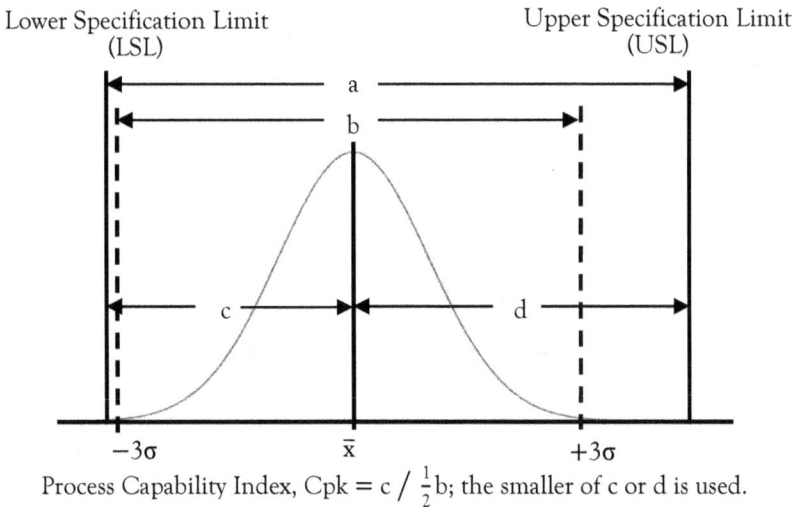

Process Capability Index, $Cpk = c / \frac{1}{2}b$; the smaller of c or d is used.

Figure 10-3: Calculation of Process Capability Index

In an off-centred process, Cp is useful when compared to Cpk as it gives an indication of the *potential* Process Capability Index when the process is centred within the specification limits.

A Process Capability Index (Cpk) less than 1 will indicate an incapable process, i.e., the process spread is greater than the specification limits. Cpk greater than 1 will indicate a capable process with confidence growing as Cpk becomes 2 or more.

Example of a Process Capability Index Analysis:

Figure 10-4 illustrates Cpk Analysis of the same data used to construct Averages and Range ($\overline{X}$ and R) Control Charts in Figure 11-9.
- Upper and Lower Specification Limits given:
  USL = 2700, LSL = 2100, and midpoint = 2400
- $Cpk = c / \frac{1}{2}b = 2262\text{-}2100 / \frac{1}{2}(6\times98.78) = 0.547$
- $Cp = a/b = 600/(6\times98.78) = 1.012$

The Cpk of 0.547, which is less than 1, indicates an incapable process. The Cp of 1.012, which is greater than 1, indicates that the process has the potential to be capable.
The process is off-centre causing the process to be incapable of yielding 100% quality; by statistical calculation, the Quality Yield = 94.95%.

Process average, $\overline{x}$ = 2262
Standard Deviation, σ = 98.78
USL = 2700
LSL = 2100
Quality Yield = 94.95%
Sample size, n = 125

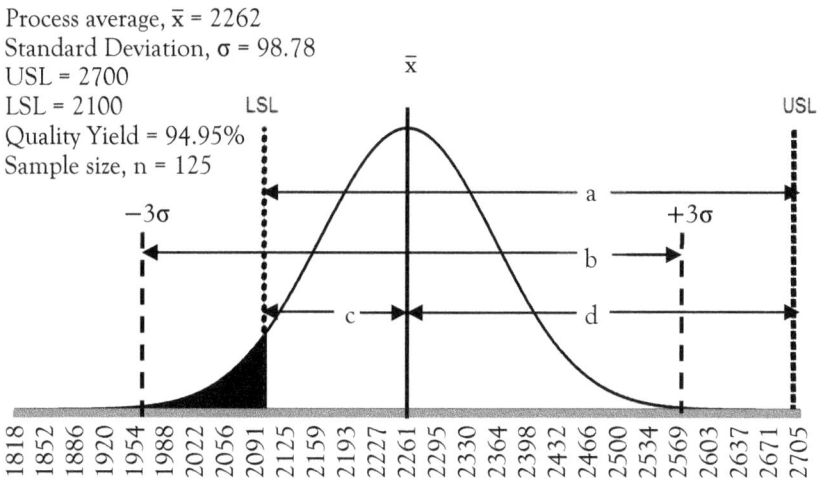

Figure 10-4: Cpk Analysis of actual data

The data used to construct the Histogram in figure 10-4 is of tensile strength taken from production batches. The cause of the off-centre distribution of data was established as being due to the heat treatment oven temperature being higher than required. This "special" cause was removed by reducing the heat treatment temperature slightly which resulted in an increase of tensile strength.

The process was centred and Quality Yield increased from 94.95% to 100%.

10.4 Capability of the product in its intended application

It is imperative for the designer or specifier to take into account the capability of the product or equipment in its intended application; the importance of considering this during the technical review, prior to contract signing, is emphasized. It would appear that this was not taken into account when viewing the results of the failure investigation study mentioned in Chapter 5.2. This study established that at least 50% of installations of conveyor chain on armoured face conveyor equipment had uncomfortably high load factors. In the cases where the load factor was excessively high, the chain was highly stressed and its life was considerably shortened, as attested to in the actual findings of the study.

It is common practice for the customer to give "nominal" quantities or values for load and stress likely to be encountered by the equipment during its operation. A "nominal" quantity or value in engineering is often given for e.g., load, stress, length, volume, voltage. This is not an exact value; a nominal quantity or value may be different from the actual value so the product or equipment designer or specifier needs to establish the actual quantity or value and the likely spread or variation of this quantity or value during operation.

Of course there are many factors that could affect the useful life of the product or equipment. The initial strength of the product would have natural variation affected by the properties of its constituent materials or components and the variations of the manufacturing processes; the capability of the product or equipment in operation could be affected by a variation in equipment configuration, operational stress or load, and operating conditions.

When various products or equipment-set building blocks, such as motors, gearboxes, couplers, drives, and controllers, are brought together to make up an equipment-set, the capability of the individual product in operation could be affected, and the use of similar but different constituent building-blocks would obviously cause a variation in equipment performance. These effects and variations must be considered by the product and equipment specifier or designer.

Operating conditions pertaining to the environment, i.e., temperature extremes and rate of change, humidity, corrosion, vibration, shock, etc., need to be considered as well as operating conditions pertaining to the man-controllable factors such as numbers of cold starts and aggressiveness of control by the work-force. The designer or specifier needs to take all "common" causes of applied stress or load into account including some "special" causes of stress or load that could arise, for instance, due to momentary shock-loading. With this in mind, the designer or specifier needs to consider a design safety margin, illustrated in figure 10-5.

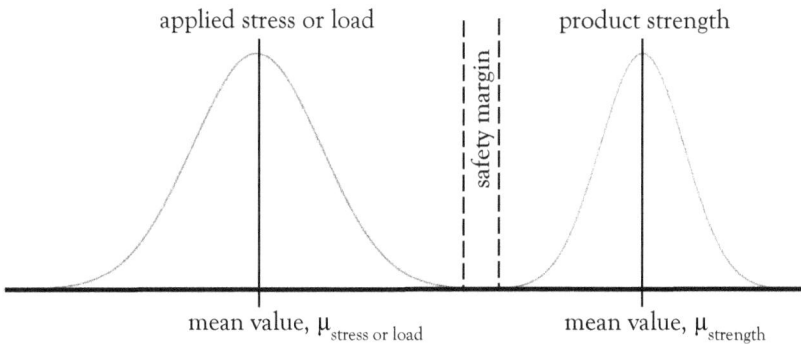

Figure 10-5: Illustration of safety margin

In figure 10-5, the natural variations of applied stress or load and of product strength are shown in normally distributed probability distribution functions having bell-shaped curves. The most frequent values occur around the mean value $\mu_{\text{stress or load}}$ and μ_{strength} and the less frequent values occur at the tails of each curve.

In figure 10-6 the probability distribution functions of applied stress or load and of product strength are in an interference state, i.e., they overlap; stress or load exceeds strength – this is the area of probable failure.

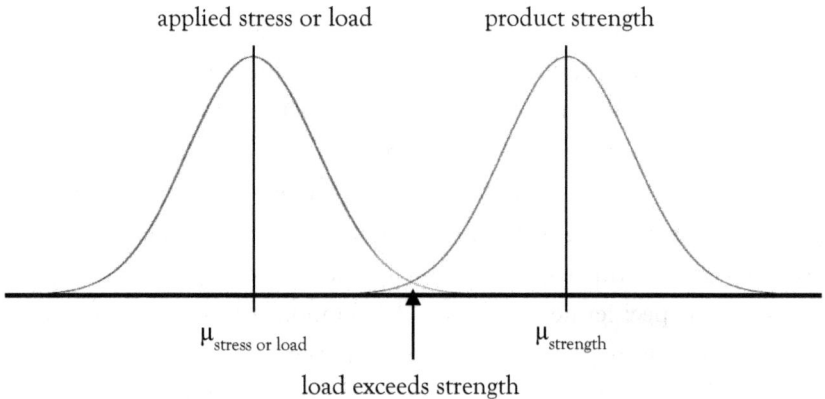

applied stress or load   product strength

$\mu_{\text{stress or load}}$ μ_{strength}

load exceeds strength

Figure 10-6 Stress or load and strength interference

Determination of the probability of failure:

Firstly we need to mathematically define the area of probable failure. Since normal distributions of stress or load and strength are assumed, the difference between the stress or load and strength distributions is also a normal distribution; this will have a probability distribution function with a mean value of $\mu_{\text{difference}}$ and standard deviation of $\sigma_{\text{difference}}$ calculated from the following equations:

(1) $\mu_{\text{difference}} = \mu_{\text{strength}} - \mu_{\text{stress or load}}$

(2) $\sigma_{\text{difference}} = \sqrt{\sigma^2_{\text{strength}} + \sigma^2_{\text{stress or load}}}$

The areas under this normal distribution (areas of probable failure) can be obtained by integrating the probability distribution function; thankfully, for convenience, a table of Standard Normal Probabilities is made available through the conversion of the normal distribution into a Standard Normal Distribution using the Z transformation.

Figure 10-7 illustrates the Standard Normal Distribution. The Z-score, also known as the standardised score, is a measure of how many standard deviations a value is away from the mean.

The Z-score of the Standard Normal Distribution for value x^i in a normal distribution with mean μ and standard deviation σ, is calculated from the following equation:

(3) $\text{Z-score} = (x^i - \mu)/\sigma$

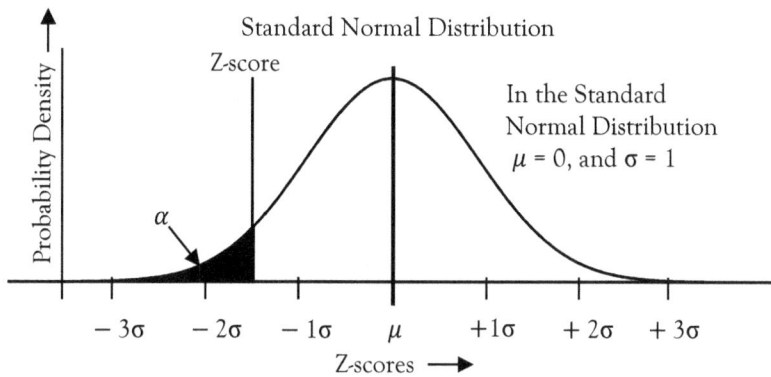

Figure 10-7: Area of probability, α, for a Z-score

The areas of probability, α, for Z-scores, can be looked up in the Table of Standard Normal Probabilities.

Table 10-2, Area under the Standard Normal Distribution curve, is an extract from this table. This extract covers the area of probability for negative Z and is provided to enable the questions in the examples that follow to be answered.

Table 10-2: Area under the Standard Normal Distribution curve

Z	\multicolumn{10}{c}{Area of probability, α for negative Z, and areas LEFT of the Z-score, extracted from table of Standard Normal Probabilities}

Z	0.00	0.01	0.02	0.03	0.04	0.05	0.06	0.07	0.08	0.09
-3.4	.0003	.0003	.0003	.0003	.0003	.0003	.0003	.0003	.0003	.0002
-3.3	.0005	.0005	.0005	.0004	.0004	.0004	.0004	.0004	.0004	.0003
-3.2	.0007	.0007	.0006	.0006	.0006	.0006	.0006	.0005	.0005	.0005
-3.1	.0010	.0009	.0009	.0009	.0008	.0008	.0008	.0008	.0007	.0007
-3.0	.0013	.0013	.0013	.0012	.0012	.0011	.0011	.0011	.0010	.0010
-2.9	.0019	.0018	.0018	.0017	.0016	.0016	.0015	.0015	.0014	.0014
-2.8	.0026	.0025	.0024	.0023	.0023	.0022	.0021	.0021	.0020	.0019
-2.7	.0035	.0034	.0033	.0032	.0031	.0030	.0029	.0028	.0027	.0026
-2.6	.0047	.0045	.0044	.0043	.0041	.0040	.0039	.0038	.0037	.0036
-2.5	.0062	.0060	.0059	.0057	.0055	.0054	.0052	.0051	.0049	.0048
-2.4	.0082	.0080	.0078	.0075	.0073	.0071	.0069	.0068	.0066	.0064
-2.3	.0107	.0104	.0102	.0099	.0096	.0094	.0091	.0089	.0087	.0084
-2.2	.0139	.0136	.0132	.0129	.0125	.0122	.0119	.0116	.0113	.0110
-2.1	.0179	.0174	.0170	.0166	.0162	.0158	.0154	.0150	.0146	.0143
-2.0	.0228	.0222	.0217	.0212	.0207	.0202	.0197	.0192	.0188	.0183
-1.9	.0287	.0281	.0274	.0268	.0262	.0256	.0250	.0244	.0239	.0233
-1.8	.0359	.0351	.0344	.0336	.0329	.0322	.0314	.0307	.0301	.0294
-1.7	.0446	.0436	.0427	.0418	.0409	.0401	.0392	.0384	.0375	.0367
-1.6	.0548	.0537	.0526	.0516	.0505	.0495	.0485	.0475	.0465	.0455
-1.5	.0668	.0655	.0643	.0630	.0618	.0606	.0594	.0582	.0571	.0559
-1.4	.0808	.0793	.0778	.0764	.0749	.0735	.0721	.0708	.0694	.0681
-1.3	.0968	.0951	.0934	.0918	.0901	.0885	.0869	.0853	.0838	.0823
-1.2	.1151	.1131	.1112	.1093	.1075	.1056	.1038	.1020	.1003	.0985
-1.1	.1357	.1335	.1314	.1292	.1271	.1251	.1230	.1210	.1190	.1170
-1.0	.1587	.1562	.1539	.1515	.1492	.1469	.1446	.1423	.1401	.1379
-0.9	.1841	.1814	.1788	.1762	.1736	.1711	.1685	.1660	.1635	.1611
-0.8	.2119	.2090	.2061	.2033	.2005	.1977	.1949	.1922	.1894	.1867
-0.7	.2420	.2389	.2358	.2327	.2296	.2266	.2236	.2206	.2177	.2148
-0.6	.2743	.2709	.2676	.2643	.2611	.2578	.2546	.2514	.2483	.2451
-0.5	.3085	.3050	.3015	.2981	.2941	.2912	.2877	.2843	.2810	.2776
-0.4	.3446	.3409	.3372	.3336	.3300	.3264	.3228	.3192	.3156	.3121
-0.3	.3821	.3783	.3745	.3707	.3669	.3632	.3594	.3557	.3520	.3483
-0.2	.4207	.4168	.4129	.4090	.4052	.4013	.3974	.3936	.3897	.3859
-0.1	.4602	.4562	.4522	.4483	.4443	.4404	.4364	.4325	.4286	.4247
-0.0	.5000	.4960	.4920	.4880	.4840	.4801	.4761	.4721	.4681	.4641

Note: The units place and the first decimal place are shown in the furthest left hand column, and the second decimal place is displayed across the top row.

Example of calculation for probability of failure:

An armoured faced conveyor (AFC) will be subjected to loads averaging 1600kN. The variation of the load is estimated to be normally distributed about a mean of 1600kN and a standard deviation of 150kN.

The analysis of manufacturing records of many batches indicates that Chain A has a normally distributed operating force of 1660kN with a standard deviation of 70kN and Chain B has a normally distributed operating force of 2170kN with a standard deviation of 90kN.

What is the probability of failure of chain A and of chain B?

AFC load: μ_{load} = 1600kN; σ_{load} = 150kN

Chain A operating force: $\mu_{strength}$ = 1660kN; $\sigma_{strength}$ = 70kN

Chain B operating force: $\mu_{strength}$ = 2170kN; $\sigma_{strength}$ = 90kN

For our purposes, $x^i = 0$

Equation	Chain A	Chain B
(1) $\mu_{difference}$	1660 – 1600 = 60	2170 – 1600 = 570
(2) $\sigma_{difference}$	$\sqrt{70^2+150^2}$ = 165.5	$\sqrt{90^2+150^2}$ = 174.9
(3) Z-score	(0 – 60)/165.5 = – 0.36	(0 – 570)/174.9 = -3.26

The areas of probability, α, for Z-score values as calculated, can be looked up in table 10.2.

From the table, for Z-score -0.36, Chain A probability of failure is 0.3594 and for Z-score -3.26, Chain B probability of failure is 0.0006

The chance of failure of Chain A is 3594 in 10000, whereas Chain B has a far smaller chance of failure of 6 in 10000.

The product or equipment having a high probability of failure will also obviously have a high probability of incurring warranty costs.

Example of calculation for probability of not meeting a declared value:

The marketing department of the manufacturer of Chain B declares that this product has a minimum operating force of 2100kN.
What percentage of produced product will with statistical probability not meet this lower operating force?
Using equation (3), Z-score = $(x^i - \mu)/\sigma$; x^i = 2100, μ = 2170, σ = 90
Z-score = (2100 – 2170)/90 = – 0.78
From table 10-2 for Z-score – 0.78, α = 0.2177

21.77% of production will have the probability of not meeting the marketed minimum operating force of 2100kN.

Aggressive environments such like those encountered in mining, and the arduous life of equipment during operation will result in sub-optimally set-up and specified equipment being quickly exposed.
- The capability of the product or equipment in operation could be affected by set-up and maintenance routines. For instance, drive-belt or conveyor chain must be correctly pre-tensioned to enable it to provide the specified performance.
- Product strength can decrease rapidly due to material fatigue occurring through cyclical stresses caused by the likes of continuous load jarring.
- It is surprisingly how rapidly stress cracking can propagate through heat friction or from corrosion pits.
- It is alarming to see how rapidly grit and, for example, coal dust, can ingress seals especially when encouraged by vibration, and cause seal-seat-surface abrasion leading to oil leaks.

Factors such as these can rapidly affect the capability and useful life of the product in its intended application, and failure of equipment within a few months of operation cause the new equipment owners to turn to the equipment supplier and expect repair or replacement under warranty. This is the situation that was often encountered in China.

Figure 10-8 illustrates what happens to product strength over a period of time; the initial strength, $\mu_{\text{strength 1}}$, decreases to $\mu_{\text{strength 2}}$ and an accumulation of wear factors cause its distribution to spread.

An interference state is reached when the strength and load distributions overlap. As this state is entered, the probability of product failure increases and progressively worsens as the product strength decreases and its distribution further widens.

Over time, the average strength decreases from $\mu_{\text{strength 1}}$ to $\mu_{\text{strength 2}}$ and the variation of strength widens

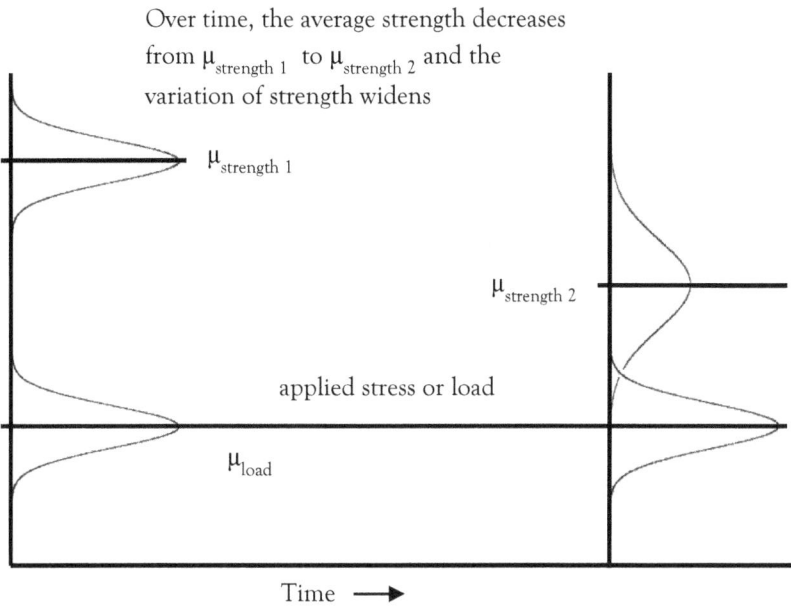

Figure 10-8: Decrease of strength over time

To maintain market-share and sustain growth, manufacturing organisations must apply continuous improvement to every facet of their business, foremost is ensuring that equipment which they produce are correctly specified for the job and manufactured to consistent quality standards. Continual attention must be given to the improvement of the capability of their products and equipment that they design and manufacture vis-à-vis intended application.

Improvement opportunities are identified through learning the capability of the product, and being able to identify the "common" and "special" causes that affect the performance of the product.

There is a great amount of learning from the collection and analysis of failure data by knowledgeable people. Information obtained in this way can be used to improve product designs and manufacturing processes. Without this, quality management is incomplete.

Questions for Chapter 10

10-1: Explain "common" and "special" causes of process variation. Give examples to support your explanation.

10-2: Explain what is meant by a "capable process" using the "Run Chart" to illustrate.

10-3: With "common" and "special" causes of variation in mind, how should process improvement be approached?

10-4: What would be the necessary steps to determine whether a process has the capability of operating within an engineering tolerance?

10-5: What actions would an organisation undertake to ensure that their products and equipment have the capability to satisfactorily perform the intended application?

11. BASIC QUALITY TOOLS

Quality tools provide the means for making decisions based on facts; they are used to aid problem identification, visualization, analysis, resolution, and quality improvement.

Seven quality tools were identified by Kaoru Ishikawa[11] which require very little training, and with which most quality-related problems can be resolved; these are referred to as Basic Quality Tools and are as follows:

1. Flow Chart or Flow Diagram
2. Check Sheet or Defect Tally Sheet (other forms of check sheets are Defect Location or Defect Concentration Diagrams and Measles Charts)
3. Histogram
4. Pareto Chart
5. Cause and Effect Diagram, also called an Ishikawa Diagram or Fishbone Diagram
6. Scatter Diagram (a Scatter Diagram with data plotted in time sequence is a Run Chart)
7. Control Chart

11.1 Flow Chart

A Flow Chart describes a process by graphically displaying the activities that occur in sequence or in parallel, showing yes/no choices, and identifying control points. This information is used to understand the workings of a process and to identify important activities.

The detail captured in a flow chart needs to be adjusted to a level that serves the purpose of helping people understand the process, and this is not accomplished if the flow chart is either too complex or too simple.

---

[11] Kaoru Ishikawa (July 13, 1915 – April 16, 1989), Japanese organisational theorist, professor, author, lecturer and quality management consultant. Kaoru Ishikawa gave a tremendous boost to Total Quality Control in China from the summer of 1978, where work began in the Beijing Internal Combustion Engine Factory. He visited China almost every year until his death.

The flow chart can be analysed and areas for improvement can be identified. Improvement could be by way of;

- Removing impediments to obtaining needed information, removing unnecessary administrative tasks, paperwork and approvals, and removing time-wasting access especially caused by people who are never available.
- Simplifying complex work flows, removing duplication, and reducing handling by combining consecutive activities.
- Eliminating many storage points of the same information, centralizing data storage, standardising reports, eliminating hard-copies and unused data.
- Identifying handover criteria and internal customer requirements.
- Reducing the process cycle time, e.g., by reducing interruption, improving physical co-location, and where possible, by doing process activities in parallel rather than sequentially.
- Standardising by selecting a single way to do an activity and having all work performers do the activity that way all the time.
- Making it difficult to do the activity incorrectly, e.g., by designing carefully thought-out forms that make it clear what information is required, and by using cross-checking.
- Evaluating activities in the organisation to determine their contribution to adding value and to meeting customer requirements.

Flow charts are logical to follow and indicate decision points very well, they also identify handover points between functions and departments; these are often the points where communication or timing problems occur.

It is common to find activities of the various quality management processes described in flow charts. These processes usually involve more than one function or department and are called deployment flow charts. Examples of deployment flow charts are: handling of non-conforming product – see figure 11-1, internal quality auditing and handling of customer complaints.

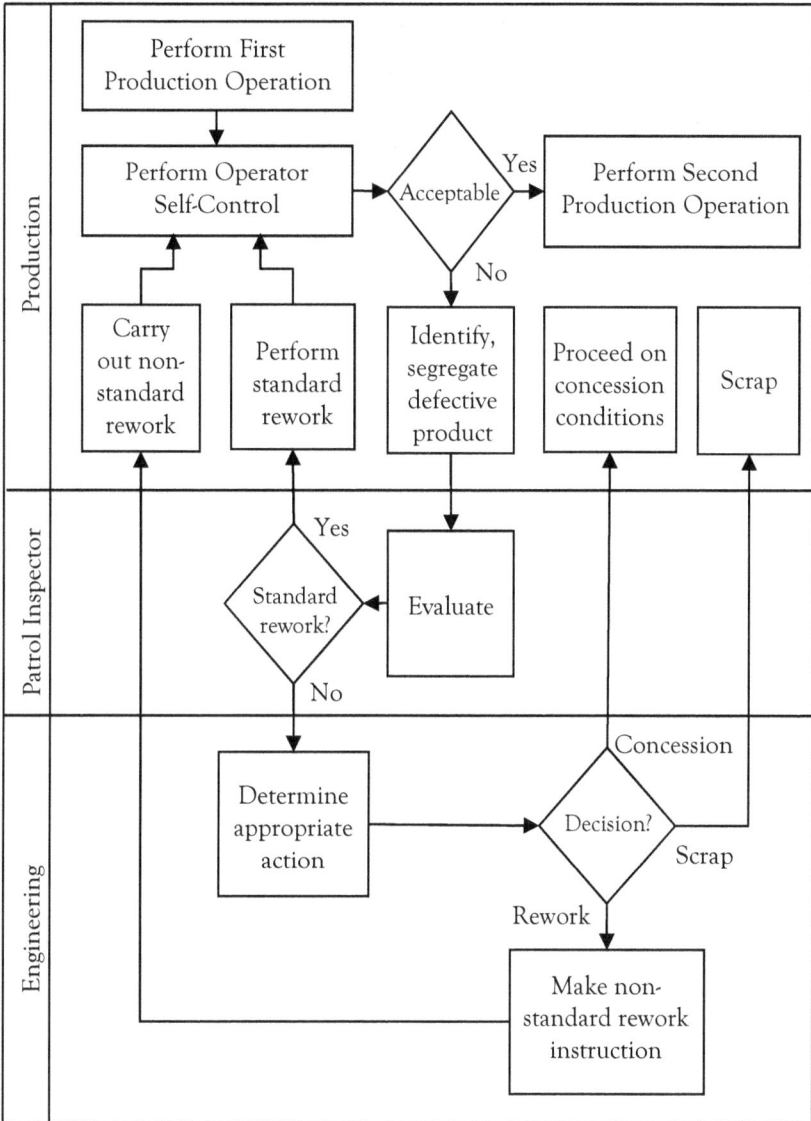

Figure 11-1: Example Deployment Flow Chart

11.2 Check Sheet

The Check Sheet is also known as a Defect Tally Sheet. This is a prepared form for collecting data and identifying the number of times that a specific cause is the source of a defect or a quality problem. Other forms of Check Sheets are Defect Location Diagrams or Measles Charts and Concentration Diagrams.

Note that a Check Sheet must not be confused with a Check List – a Check List is used to ensure important steps or actions have been taken.

Figure 11-2 illustrates an example of the use of a Check Sheet that includes a Defect Location Diagram. This Defect Location Check Sheet revealed the positional location of defects observed after the machining operation of cast aluminium housings. The information was fed back to the casting foundry where modifications to the casting process were made which practically eliminated the porosity defects.

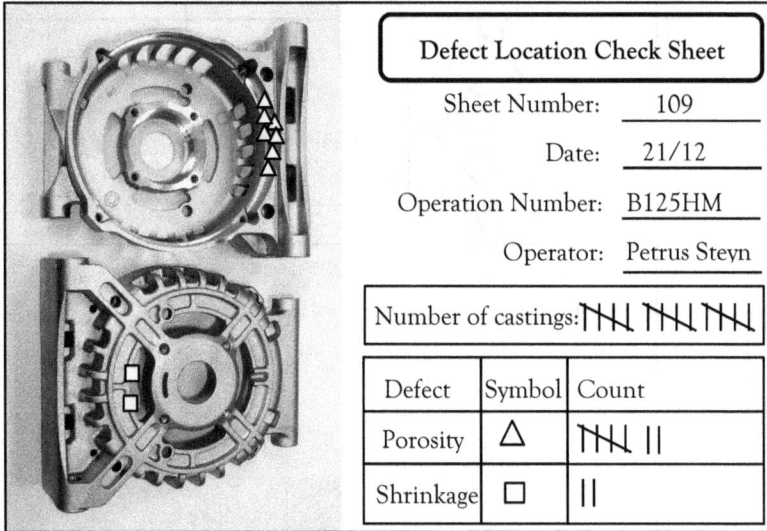

Figure 11-2: Defect Location Check Sheet

Check Sheets are effectively used in many diverse applications; to cite some instances:

- To identify the type and location of soldering defects in a printed circuit board flow-soldering process and to use this information to fine-tune the soldering process.

- To identify, in a service repair workshop, the most frequently replaced components on industrial electronic control boards and to feed this information back to the design office to improve the design of the circuit.

- To identify the area of concentration of defects in vehicle body painting and to use this information to improve the robotic painting process.

- To fine-tune applied pressure during an automotive glass lamination process.

- To determine the time periods during which the telephone receptionist received the most calls and to provide additional support during these times.

11.3 Histogram

A Histogram is a plot that provides a graphical display allowing discovery of the underlying frequency distribution (shape) and central tendency (average or mean) of a set of numerical continuous data (continuous data is information that can be measured on a continuum or scale, e.g., length, size, time, and cost).

From inspection of the plotted data, the nature and shape of the collected data can be determined and it can be seen whether the distribution is normal, skewed, bi-modal, contains outliers, etc. The basis for the application of control charts and process capability analysis is for data distribution to be normal (bell-shaped), i.e., when there is an equally likely chance of continuous data being above or below the mean.

In figure 11-3 the data collected for "Link Pitch, After Welding" is displayed in a Histogram. In this example, the data was normally distributed, but the distribution did not fall within the specification limits.

After shortening the cut length of steel bar from which the links were formed by 0.3mm, the mean of the distribution of "Link Pitch, After Welding" was corrected from 119.2mm to 118.9mm causing the distribution to fall within specification limits as shown.

Lower Specification Limit   Upper Specification Limit

Link Pitch, After Welding — Frequency of Occurrence

Freq	118.3	118.4	118.5	118.6	118.7	118.8	118.9	119.0	119.1	119.2	119.3	119.4	119.5	119.6	119.7
9								✖	✖						
8							✖	✖	✖	✖					
7							✖	✖	✖	✖	✖				
6						✖	✖	✖	✖	✖	✖				
5						✖	✖	✖	✖	✖	✖				
4						✖	✖	✖	✖	✖	✖	✖			
3					✖	✖	✖	✖	✖	✖	✖	✖			
2					✖	✖	✖	✖	✖	✖	✖	✖	✖		
1				✖	✖	✖	✖	✖	✖	✖	✖	✖	✖	✖	✖

After correction ⟹

LSL   USL

Frequency of Occurrence

Freq	118.3	118.4	118.5	118.6	118.7	118.8	118.9	119.0	119.1	119.2	119.3	119.4	119.5
9						✖	✖						
8					✖	✖	✖	✖					
7					✖	✖	✖	✖					
6					✖	✖	✖	✖	✖				
5				✖	✖	✖	✖	✖	✖				
4				✖	✖	✖	✖	✖	✖	✖			
3			✖	✖	✖	✖	✖	✖	✖	✖			
2		✖	✖	✖	✖	✖	✖	✖	✖	✖			
1	✖	✖	✖	✖	✖	✖	✖	✖	✖	✖	✖		

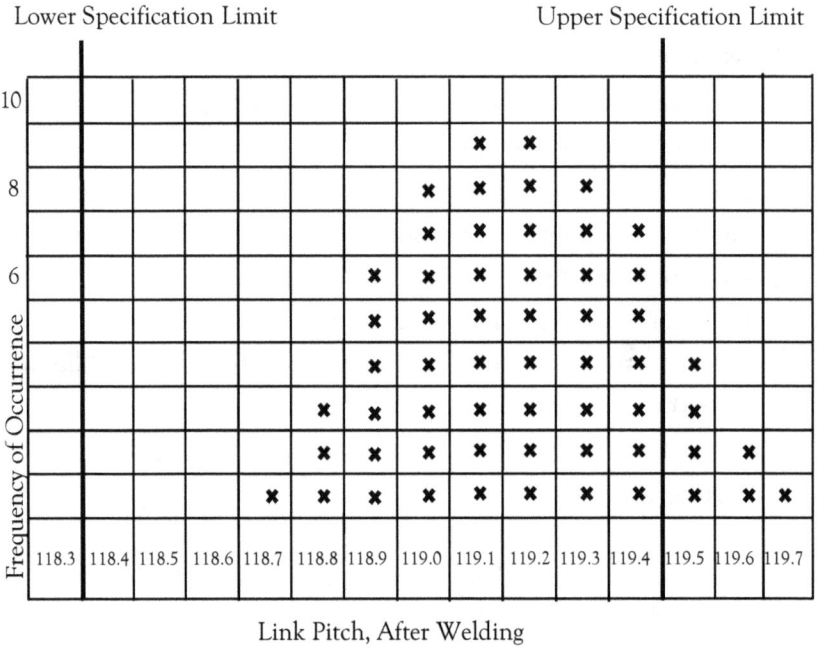

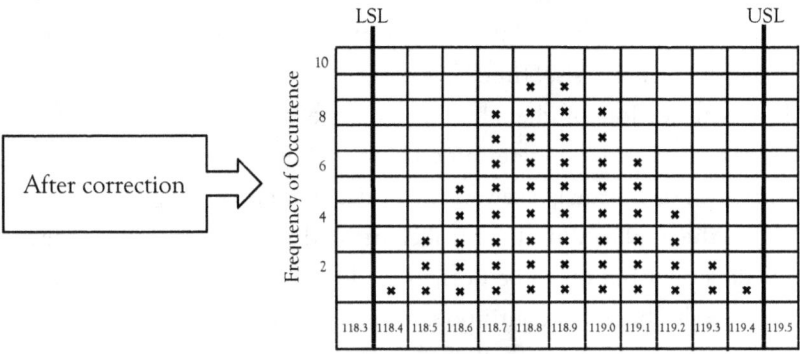

Figure 11-3: Histogram

11.4 Pareto Chart

A Pareto Chart is a vertical bar chart that puts data such as causes of defects or types of quality problems in a descending order of popularity or frequency of occurrence; see figure 11-4. The data from the Check Sheet can be used to create a Pareto Chart.

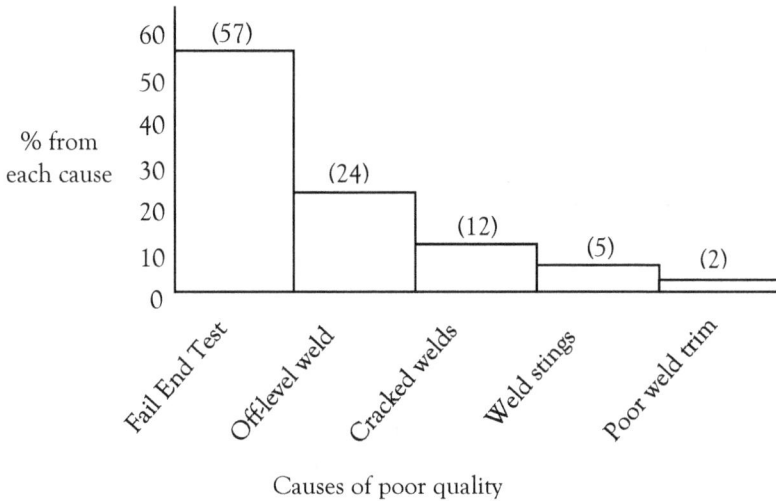

Figure 11-4: Pareto Chart or Diagram

Pareto Charts are used to prioritize specific defects or quality problems and also to verify that implemented improvement continues to work.

11.5 Cause and Effect Diagram

The Cause and Effect Diagram is also called an Ishikawa Diagram as well as a Fishbone Diagram.

The problem statement is the effect and the possible causes are shown as leading to, or potentially leading to the effect. The major areas of possible causes are shown as the main bones of the diagram. Causes are systematically eliminated to find the root cause.

An example of a Cause and Effect Diagram is provided in figure 11-5. In this example, the problem being investigated was "Low Ultimate Strength of Forged Link Legs". The problem was identified when product proving results were being studied during the Production Readiness Design Review stage.

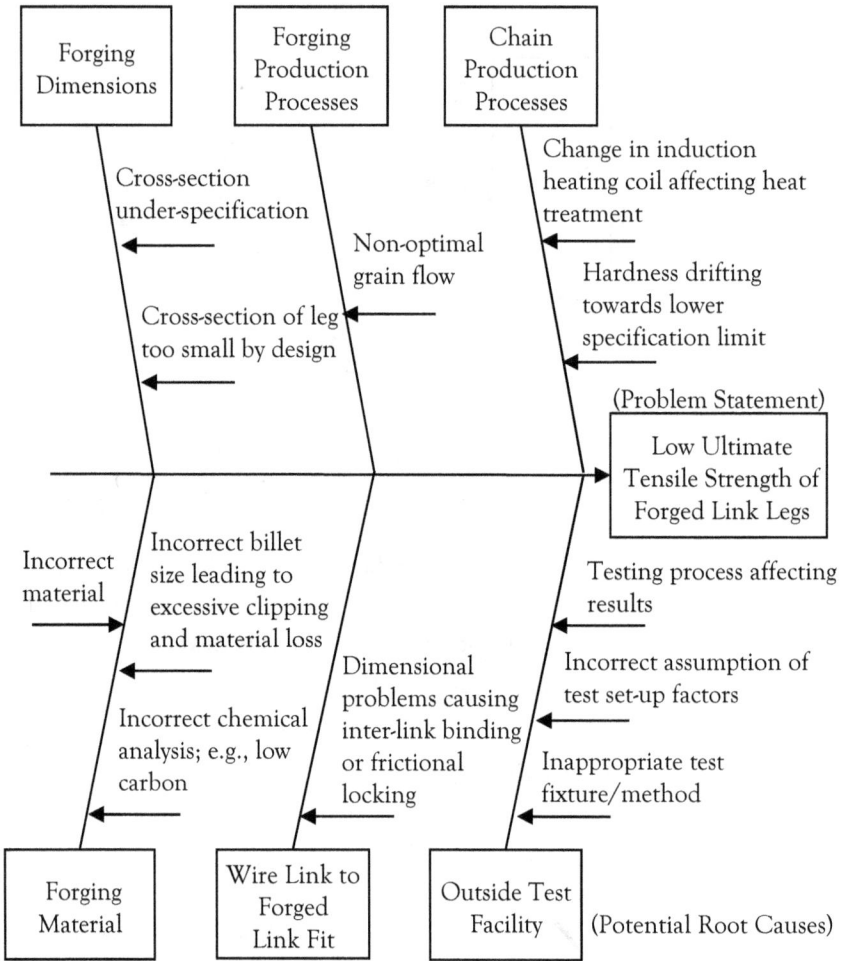

Figure 11-5: Cause and Effect Diagram

The problem initially baffled the technical experts but, after the potential causes were identified on a Cause and Effect Diagram and systematically eliminated, the cause of the problem was soon located. This was found to be "Testing process affecting results". The tensile testing machine in the approved Outside Test Facility could not provide a sufficiently high tensile test force, so the Outside Test Facility had sent the chain to another facility. The tensile testing machine at this facility applied the progressive tensile force in a non-linear step-manner, causing shock loading during step transition.

11.6 Scatter Diagram or Run Chart

A Scatter Diagram shows the relationship of one variable to another and makes it easy to find trends and correlations between the two variables. The horizontal (x) axis represents measurement values of one variable and the vertical (y) axis represents measurement values of the second variable. If one of the variables can be controlled by the examiner, it is called the control parameter and is customarily plotted along the horizontal (x) axis. The dependant variable is plotted along the vertical axis.

A Scatter Diagram that displays data in a time sequence is a Run Chart. See figure 11-6. Process data monitored on a Run Chart can reveal evidence of "special" cause variation.

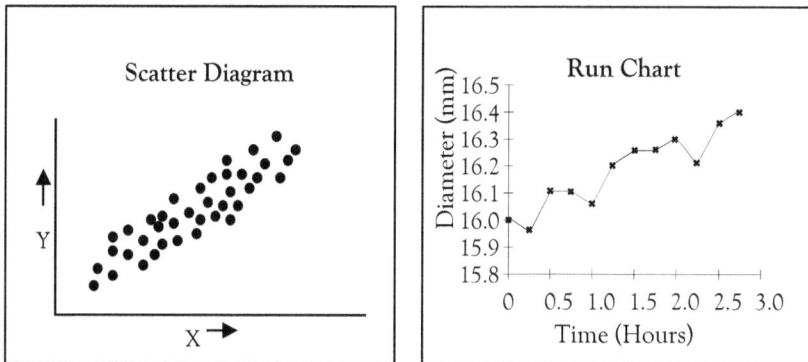

Figure 11-6: Scatter Diagram and a Run Chart

11.7 Control Chart

Control charts are used to communicate quality performance and to draw attention to when actions are needed to improve quality. When used to monitor a manufacturing process, control charts can indicate trends and draw attention to when a process is out of control. Control Charts are also called Shewhart Control Charts after their originator, Dr. Shewhart[12].

A Control Chart displays statistically determined upper and lower control limits (UCL and LCL) drawn on either side of a process average or mean; see figure 11-7. The UCL and LCL are previously determined through statistical calculations of data from earlier trials. The Control Chart shows whether the collected data of process values are within UCL and LCL.

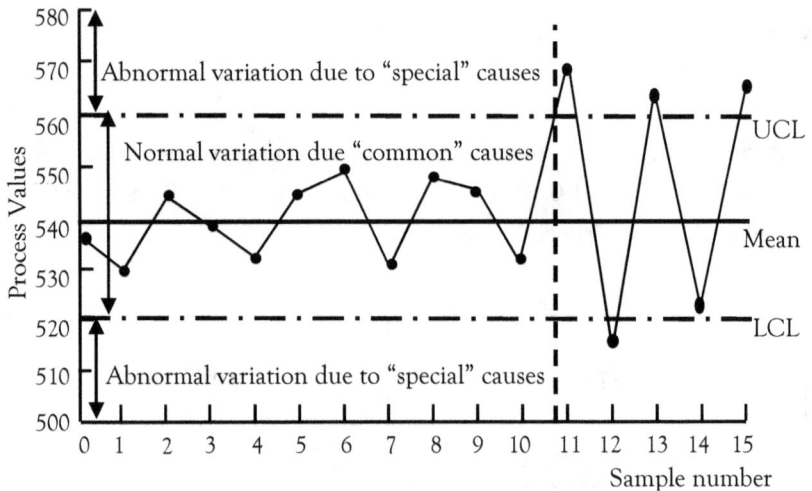

Figure 11-7: Control Chart

When a process is operating in a stable and predictable state, small process variation occurs between UCL and LCL; this variation is said to arise from "common" causes inherent to the process. This stable and predictable state is reflected to the left of the vertical dotted line in

12 Walter Andrew Shewhart: March 18, 1891 - March 11, 1967 was an American physicist, engineer and statistician. Two of his contributions continue to influence the daily work of quality, i.e., control charts and the Plan–Do–Study–Act (PDSA) cycle, popularized by W. Edwards Deming as Plan–Do–Check –Act (PDCA) cycle.

figure 11-7. To the right of the dotted line, the process variation moves to an out-of-control state. This is due to assignable causes, also referred to as "special" causes. Examples of some causes of variation due to "common" and "special" causes are given in figure 10-2.

"Variable Control Charts" are used with quality characteristics that can be measured and expressed numerically, e.g., temperature, strength, diameter and thickness. Probably the most popular is the Averages and Range Control Chart ($\overline{X}$–R Chart). Because of the usefulness of the $\overline{X}$–R Chart, an explanation of the principle of control limits, and construction and interpretation of these charts immediately follow in this chapter.

"Attribute Control Charts" are used with quality characteristics that cannot be expressed numerically, or when it is impractical to do so. Items are classified into go and no-go, defective and non-defective, or conforming and non-conforming.
C, U, p and np Charts are commonly used for controlling attributes.
- C Chart for number of defectives per production run, or per batch, or per day, etc.
- U Chart for rate of defectives, i.e., the number of defectives divided by the number of units inspected
- p Chart for proportion or fraction of defectives per batch, or per day, etc.
- np Chart which is similar to the p Chart but may be easier to understand because it plots the number of non-conforming units instead of the proportion of non-conforming units.

Averages and Range Control Chart ($\overline{X}$–R Chart):

An $\overline{X}$–R Chart comprises two charts, the process mean and the process range. The process mean is plotted on an $\overline{X}$ Chart and process range on an R Chart over time in sub-groups of a predefined size. The $\overline{X}$ Chart and the R Chart are displayed together because both charts must be interpreted to determine whether the process is stable.

These control charts present a visual picture of process performance. They are sensitive to process variation and their statistically derived control limits provide a greatly improved probability of detecting out-of-limit measurements; for instance, on an Averages or $\overline{X}$ Chart this probability is of the order of 80% for averages of a sample of five, as opposed to that of a Run Chart's single low digit % based on individual measurements of one. This greatly improved probability enables an early detection of problems so that immediate corrective action can be taken to prevent further out-of-limit measurements.

The $\overline{X}$-R Chart enables the process to be controlled to its optimum performance.

It must be noted that it is inappropriate to include specification limits on a Control Chart. The values on an $\overline{X}$ Chart are determined from the average of a sample of measurements and are not individual measurements. Specification limits are individual measurements and the frequency distribution curve for averages of samples will have a smaller standard deviation than the frequency distribution curve for individual measurements, as shown in figure 11-8.

Principle of Control Limits:

With the aid of figure 11-8 the principle of control limits and an understanding of the relationship of control limits and specification limits are explained.

The distribution curve (1) is for individual measurements of one; this curve is shown in Position-1 touching the Upper Specification Limit. Inside this curve is distribution curve (2) which is formed for illustration purposes by taking averages of samples of five.

Because curve (2) is plotted from the same information as curve (1), curve (1) and (2) have the same mean and curve (2) is always located inside of curve (1) as shown. Movement up or down of the curve of individual measurements of one will result in movement up or down of the curve for averages of samples of five.

When curve (1) is just touching the Upper Specification Limit, a line is drawn 1.96 times the standard deviation for samples of five, or, for other sample sizes, $1.96\sigma_n$ above the mean of curve (2). This defines the position of the Upper Control Limit, as shown. A Lower Control Limit is established in the same way based on the Lower Specification Limit.

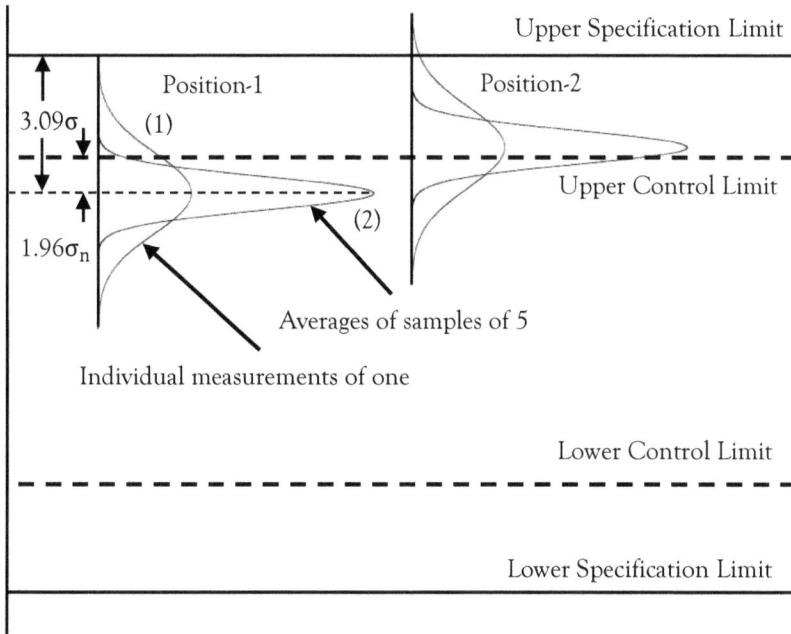

For normal distribution (1), 3.09 is the approximate value of the 100 percentile point, and within approximately ±3.09 standard deviations (σ) of a normal distribution's mean lays 100% of its area.

For normal distribution (2), 1.96 is the approximate value of the 97.5 percentile point and within approximately ±1.96 standard deviations of a normal distribution's mean lays 95% of its area.

Figure 11-8: The principle of control limits

When the process is in Position-1, most of the averages of samples of five will be within the Control Limits. The exception will be if a sample of five is drawn which, when averaged, comes from the upper tail of curve (2) which is above the Upper Control Limit in the illustration. The probability that this will happen is one in forty (2.5/100) and will be a warning that the process is just about to go out-of-limit and so it should be reset.

If after some time the process moves to Position-2 where samples are measured and found to exceed the Upper Specification Limit, represented in the illustration by the upper tail of the curve for individual measurements being just outside the Upper Specification Limit, notice that most of curve for the averages of samples of five is outside the Upper Control Limit. This means that there is a huge probability of detecting this shift on the average of a sample of five, and this probability increases as the sample size increases.

Construction of Averages and Range Control Charts:

Equations used to construct the Averages and Range ($\overline{X}$ –R) Control Charts (Read $\overline{X}$ as "X bar", $\overline{\overline{X}}$ as "X bar-bar", $\overline{R}$ as "R bar"):

(1) Average for sub-group;

$\overline{X}$ = sum of sub-group measurements/sub-group size (n)

(2) Range for sub-group;

R = Largest in sub-group – Smallest in sub-group

(3) Average of Averages;

$\overline{\overline{X}}$= sum of sub-group averages/number of sub-groups

(4) Average of Ranges;

$\overline{R}$= sum of sub-group ranges/number of sub-groups

(5) $\overline{X}$ Chart UCL = $\overline{\overline{X}}$ + (A$_2$ × $\overline{R}$), can also use $\overline{\overline{X}}$ + $3(\sigma/\sqrt{n})$

(6) $\overline{X}$ Chart LCL = $\overline{\overline{X}}$ – (A$_2$ × $\overline{R}$), can also use $\overline{\overline{X}}$ – $3(\sigma/\sqrt{n})$

(σ is the population standard deviation of individual measurements; σ_n is the sample or sub-group standard deviation for size n, and σ_n = $\sigma/\sqrt{n}$)

(7) R Chart UCL = $\overline{R}$ × D$_4$

(8) R Chart LCL = $\overline{R}$ × D$_3$

(9) Table 11-1 contains $\overline{X}$ –R Control Chart Constants A$_2$, D$_3$ and D$_4$ for sub-group (n) sizes of 4, 5 and 6.

Table 11-1: Control Chart Constants (from ASTM: MNL 7-8TH)

Sub-group size (n)	A_2	D_3	D_4
4	0,729	–	2.282
5	0.577	–	2.114
6	0.483	–	2.004

Note: Sample or sub-group sizes of 4, 5 and 6 are practical and most frequently used.

The Control Chart sensitivity to detecting process variation increases as the sub-group size increases due to the standard deviation of the distribution of averages (σ_n) decreasing, making the control limits become tighter ($1.96\sigma_n$ in figure 11-8 becomes smaller). From a practical perspective a manageable sub-group size is specified that will allow the appropriate sensitivity for shifts in process variation to be detected with high probability.

Measurements and calculations used in the example $\overline{X}$ and R Control Charts, figure 11-9:

- A sample sub-group size (n) of five was used.

  Every hour, for each process point, five consecutive samples were taken and measured.

  Process point 1 (sub-group 1): 2275, 2268, 2221, 2537, 2213

  $\overline{X}$ sub-group 1 = (2275+2268+2221+2537+2213) / 5 = 2302.8   (1)

  R sub-group 1 = 2537 – 2213 = 324   (2)

- This was repeated for the remaining process points, i.e., sub-groups 2 to 25

- The sub-group averages were averaged together to give the Average of Averages, $\overline{\overline{X}}$, and the Average of Ranges, $\overline{R}$:

  Average of Averages, $\overline{\overline{X}}$ = 2262   (3)

  Average of Ranges, $\overline{R}$ = 233   (4)

- Control Limits were calculated:

  $\overline{X}$ Chart UCL = 2262 +(0.577 × 233) = 2396.4   (5)

  $\overline{X}$ Chart LCL = 2262 – (0.577 × 233) = 2127.6   (6)

  R Chart UCL = 233 × 2.114 = 492.6   (7)

  R Chart LCL = 233 × 0 = 0   (8)

- The $\overline{X}$ and R Control Charts were constructed:

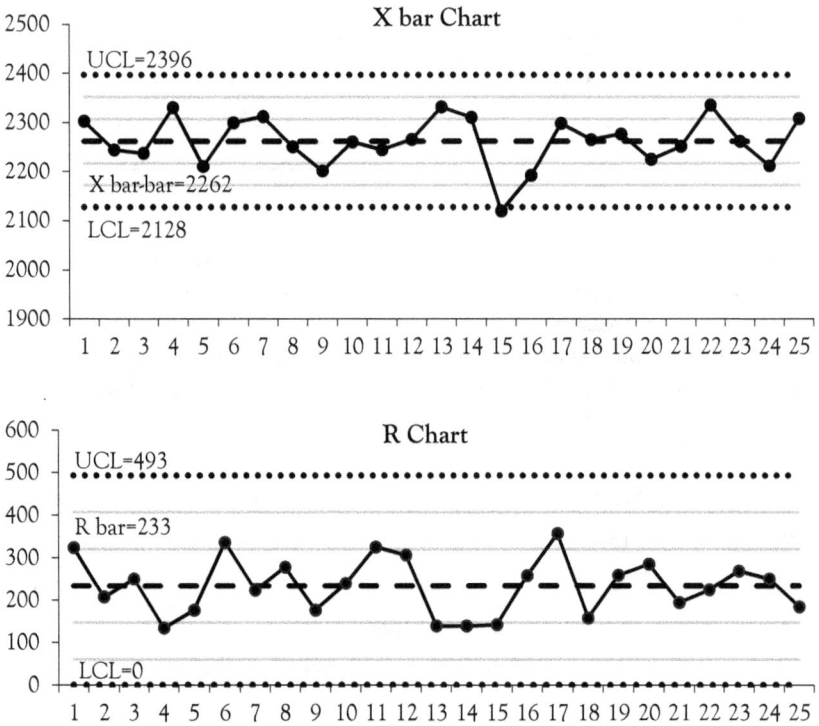

Figure 11-9: Example Averages and Range Charts

Interpretation of $\overline{X}$ and R Charts:

The $\overline{X}$ Chart focuses attention on the average and is good for detecting out-of-limit situations such as process point 15 in figure 11-9; on the R Chart the plotted value of process point 15 is relatively small and within the R Chart control limits which shows that the variation of measurements in point 15 is low. On the $\overline{X}$ Chart, values of process points 16, 17, and onwards lie within the control limits indicating that point 15 could have arisen due to a "special" cause such as random imperfections in the batch of raw material, or that the problem (e.g., tool breakage) was promptly detected and eliminated by the operator, causing the $\overline{X}$ Chart to return to statistical process control.

The R Chart is designed for detecting changes in variability, and figure 11-10 illustrates why $\overline{X}$ and R Charts are usually drawn together.

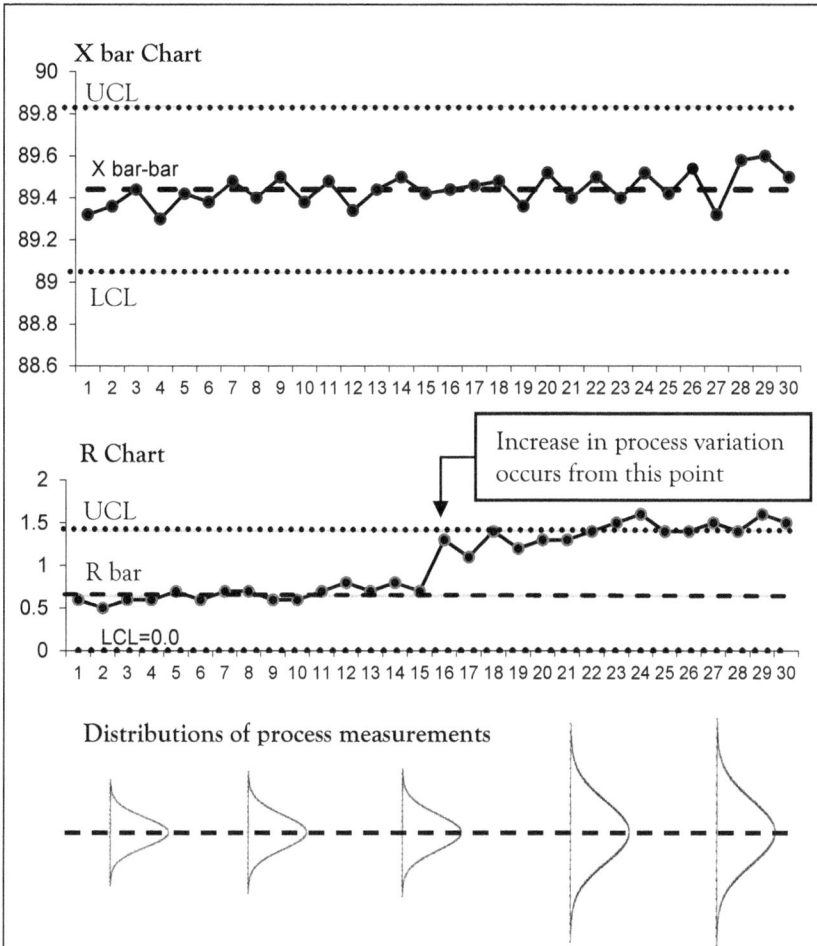

Figure 11-10: Increase in process variation detected by an R Chart

In figure 11-10, each data point on the $\overline{X}$ Chart can be visualized as containing a spread in accordance with its data point on the R Chart; the distributions of process measurements corresponding to selected R Chart data points is illustrated in the lower picture. Both $\overline{X}$ and R Charts indicate that the process is in statistical control up to the point where the process variation increases, as indicated on the R Chart. From this point onwards, the plotted points on the R Chart straddle the Upper Control Limit of the R Chart; this sends the signal that there is excessive variability around the target requirement. The problem in this instance could be, for example, a widely fluctuating temperature setting of a furnace. Improvement efforts must be focused firstly on removing such "special" causes of variability.

Note that when the process variation on the R Chart moves beyond the chart's Upper Control Limit, the process is deemed to be not in statistical control, regardless of what the $\overline{X}$ Chart indicates. This is the case from process point 23 on the R Chart in figure 11-10. When the plotted points on the R Chart occur around or below the mean Range (Average of Ranges, $\overline{R}$), indicating low variability in measurements, the position of plotted points on the $\overline{X}$ Chart and possible patterns and trends on this chart would need to be considered.

Tolerance that can be maintained by an in-control machine or process:

The question sometimes arises of what tolerance a certain machine or process could maintain with a good degree of consistency. In order to determine the answer, Range data is required because there is a relationship between Range, machine or process variation and standard deviation. A single Range measurement will provide only the very roughest of indications on variability, but the average of many Ranges will provide a good indication. If historical data of the machine or process variation in doing a similar operation is not available, it will be necessary to obtain this variation though measuring sub-group Ranges.

At least ten, preferably twenty or more sub-group Ranges would be needed to calculate $\overline{R}$, and to ensure accurate results, the machine or process variation should be checked to arise from "common" causes, as previously discussed, and any "outliers" should be discarded.

To convert the mean Range ($\overline{R}$), to an estimated standard deviation ($\sigma_{estimated}$ written σ_n), $\overline{R}$ must be multiplied with a constant; the value of this constant depends on the sub-group size given in table 11-2.

Table 11-2: Constants for converting mean Range to estimated standard deviation (from ASTM: MNL 7-8TH)

Sub-group size (n)	4	5	6	7
Constant	0.49	0.43	0.40	0.37

m ± 3 × ($\overline{R}$ × applicable sub-group constant) or m ± $3\sigma_n$ will give the estimated tolerance that can to be maintained. The midpoint value (m) of the tolerance must be able to be accommodated by the machine.

Example of estimating the tolerance that can be maintained:

Given a tolerance midpoint, m = 2400, and using information in figure 11-9, where $\overline{R}$ = 233, and a sub-group size of 5 was used, what tolerance could be maintained with good consistency by the machine in question?

- The Average of Ranges, $\overline{R}$, is required to be converted to the corresponding estimated standard deviation (σ_n)
  From table 11-2 the sub-group constant for n = 5 is 0.43
  Converting $\overline{R}$ to the corresponding standard deviation:
  σ_n = $\overline{R}$ × applicable sub-group constant = 233 × 0.43 = 100.19
- Given midpoint, m = 2400; from m ± $3\sigma_n$ the machine or process in question will be able to control to a tolerance having an Upper Specification Limit of 2700 and a Lower Specification Limit of 2100.

Questions for Chapter 11

11-1: Mark a cross in the block describing the use of each quality tool.

	Understand processes	Problem identification	Problem visualization	Problem analysis
Flow Chart				
Check Sheet				
Histogram				
Pareto Chart				
Run Chart				
Control Chart				
Cause and Effect Diagram				

11-2: (a) Briefly explain the purpose and operation of a Control Chart.

(b) Describe the Averages and Range Control Chart ($\overline{X}$ -R Control Chart) and explain its special abilities.

11-3: Construct an $\overline{X}$-R Chart from data in the table below.

Subgroup	Process values taken every hour					$\overline{X}$	R
1	89.2	89.4	89.2	89.5	89.2		
2	89.5	89.2	89.4	89.5	89.1		
3	89.4	89.6	89.5	89.3	89.3		
4	89.3	89.2	89.5	89.4	89.1		
5	89.5	89.3	89.4	89.7	89.3		
6	89.3	89.7	89.4	89.2	89.2		
7	89.2	89.5	89.4	89.6	89.7		
8	89.2	89.6	89.6	89.3	89.1		
9	89.6	89.3	89.6	89.2	89.8		
10	89.6	89.7	89.4	89.1	89.1		
11	89.2	89.9	89.6	89.3	89.4		
12	89.8	89.4	89.3	89.2	89.1		
13	89.5	89.6	89.0	89.4	89.7		
14	89.5	89.1	89.9	89.4	89.6		
15	89.2	89.9	89.5	89.3	89.1		

11-4: (a) Can the machine with the performance results in the previous question maintain an engineering tolerance of 89.5 ± 0.7mm?

(b) Comment on issues that need to be observed and actions that could be taken to improve the capability of the machine.

ANSWERS TO QUESTIONS

Answers to Questions: Chapter 1

1-1

In the table below, alongside (a) characteristics and understandings shared by organisations that have persevered and developed successful quality management practices of very high standards, is an explanation of (b) how or why these characteristics and understandings support quality management and business success.

(a) Characteristics and understandings	(b) Explanation of how or why these support quality management and business success
1. Quality is of key importance to a successful business. (See definition of quality in Appendix 1.)	Customers are the lifeblood of the business and they have quality needs and expectations which span the entire contact and relationship with the business. Understanding that these needs and expectations are important, and meeting them, triggers success.
2. Key purposes of the quality management system (QMS): - to provide confidence in the ability to meet customer requirements; - to serve as a quality problem preventive tool; - to provide a cost or differential based business advantage.	- Providing confidence in the ability to meet customer requirements is fundamentally important to business success. - Quality problem prevention leads to better business effectiveness and efficiency. - Preventing problems saves money; controlling to customer requirements creates customer focus.
3. Quality management is fully supported and committed to by top management.	Top management's actions include ensuring the everyday application of the QMS, the attainment of good quality and of ongoing quality improvement. These actions support business success.

4. Essential in sustaining quality-orientated behaviour: - Pro-quality actions of managers and supervisors. - Empowerment of employees. - Employee recognition.	- Quality is a priority of managers and supervisors and they lead by example and give support to employees. - Employee empowerment starts with training that enables them to "think quality", and is encouraged when they are allowed to make decisions affecting quality without fear, ridicule, or retribution. - Ongoing quality-orientated behaviour is reinforced and promoted through recognition and reward.
5. "Values and norms" are espoused and practiced that influences employees to have a quality-orientated behaviour.	Values and norms shape the behaviour of an organisation's employees; values and norms that support quality actions are emphasized.
6. Processes are understood to the extent that their performance can be optimised.	When processes are well understood through careful observation, measurement and analysis, then they can be optimised for better business effectiveness and efficiency.
7. Implementation of effective quality management is accelerated, and gains are most successfully held when driven by senior-level, able, assertive, tenacious, and quality-passionate "quality champions".	The many obstacles to overcome in the implementation of effective quality management are mostly management related, often requiring inter-department co-operation, and sometimes are resource intensive. Senior-level "quality champions" have great influence and impact in bringing about effective implemented solutions.
8. Facts, evidences and accurate data analysis are used for decision-making.	The effectiveness of quality management (including quality control and improvement), a) relies on using facts, evidences and accurate data analysis to make decisions, b) and not on hunches and guesses that often lead to a waste of business resources.
9. Relationships with suppliers and sub-contractors require	The business is dependent on the quality, cost and delivery performance of its suppliers. This criterion is continually being

constant management to optimise quality, cost and delivery performance.	influenced by internal and external factors, and therefore the relationship of a business with its suppliers needs ongoing management.
10. Management of quality and quality improvement requires: - constant work;	- The management of quality must receive constant attention to ensure that the various and ever-changing customers' needs and business operational issues and variances are taken care of, and quality failure and the associated failure costs are avoided.
- the continual involvement of all employees;	- Everybody in the organisation is involved in contributing to business success, and quality management is entwined with the business operations. There cannot be non-participants in a truly quality-orientated business organisation; nobody is excluded.
- quality education of all employees	- Quality education imparts a common language and meaning of quality to employees, and teaches them how to prevent problems and control quality.

1-2

The following could be possible factors or detracting influences in organisations in China that impedes the realisation of the true intent of quality management and continuous improvement:

1. Implementation of quality management processes could be impaired if the intent behind the various processes is not understood.

2. Quality management processes are not practised as the routine way of working or not accepted by all concerned to be the routine way of working.

3. Lack of management commitment and indifferent management support in ensuring that quality management processes are

effectively implemented will result in the lack of operational benefit being realised from quality management processes.

4. Not understanding that the processes of quality management are all geared towards operational effectiveness.

5. Lack of employee empowerment and a work environment that does little or nothing to encourage the use of initiative to participate in quality problem prevention, quality control, and continuous improvement.

6. Reluctance of employees to point out quality problems and to take action.

7. Not understanding risk-based thinking; not using quality management processes of contract review, design review, and management review to identify, mitigate and, wherever possible, eliminate quality risk to the organisation – i.e., eliminate the risks of the product or service not satisfying stated or implied needs.

8. Not determining a balanced set of quality objectives that covers all areas of the organisation (besides production). Also, not deploying quality objectives to relevant departments and levels in the organisation.

9. Inadequate collection of quality information (from e.g., monitoring and measuring, and internal quality audit) and unsatisfactory use of it for corrective and improvement action.

10. Not understanding that, bedsides the quality management process of preventive action, there are other quality management processes geared to avoid or prevent quality problems from occurring in the first place; these are such as quality planning, design review, quality awareness education, in-process verification, and internal quality audit.

11. Resistance of some managers to accepting certain quality management principles.

12. Lack of continual training in modern quality management concepts and practices, and in their adaptation for practical implementation.

Answers to Questions: Chapter 2

2-1

Summarized below are practices, values, beliefs, customs, work styles and relationships that define organisational or corporate culture and which affect behaviour, motivation and performance of an organisation's members:

1. The way employees are treated, i.e., as valued individuals.
2. How freely information flows through the organisational structure.
3. The freedom allowed in decision-making.
4. How initiatives, new ideas and improvements are encouraged.
5. The motivation given to employees for achieving the goals of the organisation instead of their own goals.
6. The style of management and leadership, e.g., paternalistic or Asian-paternalistic, authoritative, directive/coercive, autocratic, democratic/participative/consultative, coaching, pace-setting, persuasive
7. Duty and obedience.
8. The importance of relationships, saving "face" and preserving harmony.
9. The things that top management and leaders consistently pay attention to and emphasize, measure and control.
10. The importance in value to the organisation that top management place on quality-related actions as demonstrated in their follow-through to ensure that these actions are being carried out, and in their taking of interest in the effectiveness of the result.
11. The insistence by management that corrective, preventive, and improvement action is taken at every level in the organisation.
12. The outlook that the organisation's top management has on new initiatives and demands.

2-2

The ten values of a quality management culture described in Chapter 2 enable and support aspirations of a business organisation such as the securing and maintaining of customer satisfaction, the obtaining of a high level of efficiency and effectiveness of internal operations, and the achieving of continuous improvement to processes, products and services, in the following manner:

The organisation can only be effective if it focuses on the customer and what the customer wants (value 1). This must begin with understanding the needs and expectations of the customer, and satisfaction must be maintained by ensuring that these needs and expectations are met. Also, the customer's needs and expectations are likely to change over a period of time, and competition in the market-place is ever-present. This means that the business organisation must maintain vigilance and listen to their customers. If the organisation loses sight of this and of the key performance indicators that drive customer satisfaction (value 4), many of which should address the performance of the organisation – e.g., performance against agreed delivery schedule, time to handle customer issues, reduction of the cost of poor quality, elimination of reliability affecting design weaknesses – its effectiveness as a business is jeopardised.

Optimal performance (efficiency and effectiveness) of the process-chain within the organisation is made possible when internal suppliers know and support their internal customers, and know and meet the requirements of their internal customers (value 2). An environment that encourages teamwork and co-operation (value 3) increases the efficacy of the process-chain in "getting the job done right the first time".

From time-to-time various problems are bound to crop up in the process-chain, in the quality conformance of products, and in general; progress in resolving these are more effectively forged by focusing on finding solutions (value 7) and through teamwork. This is best achieved in the absence of finger-pointing or blame. Progress towards finding solutions to problems and to better quality management is much more effective and accelerated when there is a focus on facts (value 6) and on finding the root-cause of the problem (value 5), thus avoiding repeating the issues that cause ineffectiveness in the organisation's internal processes.

Quality management really becomes effective when it is integrated with overall management (value 9) to become a normal work routine. With this integration and involvement of everyone (value 8), the intent of quality management to prevent problems and to quickly detect issues that detract from the effectiveness of internal operations in achieving business success, is realised.

It goes without saying that the value that employees bring to every activity of the organisation counts towards the success of the organisation (value 10). When employees take responsibility for their own actions and check their own activities, effectiveness and efficiency in the organisation is increased. When employees are able to participate in the improvement process, problems are quickly detected and resolved – job fulfilment and satisfaction results, and overall performance achievement is increased. Importance of the effective actions of employees is acknowledged and their contribution to quality achievement is encouraged through recognition.

2-3

Possible actions that could oppose each of the ten principle values of the quality management culture discussed in Chapter 2:

Actions that oppose Value 1 – Focus on the customer:
- Slow in responding to customer queries.
- Not using customer-related information, as well as feedback from the customer, to improve products, delivery and services.
- Non customer-facing managers in the process-chain not focused on the customer.
- Not recognizing employees for exceptional quality related actions and service to customers.

Actions that oppose Value 2 – Know your internal customers and their requirements, and support them:
- The internal supplier not actively identifying and striving to fulfill their internal customer's requirements.
- "Throwing the file over the wall", i.e., passing on information without providing explanation or support.
- Internal suppliers not responding with appropriate urgency to issues and problems concerning products and services supplied by them.

Actions that oppose Value 3 – There must be teamwork and co-operation:
- "Power-play" and creating boundaries that inhibit co-operation and the flow of information.
- Not asking for, or accepting advice, to help achieve the desired outcome.
- Covering up issues rather than exposing them.

Actions that oppose Value 4 – Customer satisfaction must drive key performance indicators:
Not establishing customer satisfaction key performance indicators for metrics such as;
- response time to answer customer queries
- performance against agreed delivery schedule
- reduction in the number of customer complaints
- time to handle customer complaints to satisfactory resolution
- growth of number of customers
- elimination of a specific reliability affecting design weakness

Actions that oppose Value 5 – Concentrate on finding the root cause of the problem:
- Only interested in "quick-fixes".
- Not continually conducting analyses of data and examining for problem repetition, patterns and trends.
- Not following through to ensure corrective action is consistently effective.
- Allowing prejudice to affect problem identification.

- Believing that the PDCA methodology for improvement can only be applied by certain skilled employees.

Actions that oppose Value 6 – Quality management must be fact-based:
- Management not insisting on quality metrics to obtain a factual view of the measure of quality in the organisation.
- Using guesswork to direct problem-solving actions.
- Not training employees to collect and analyse and interpret data.

Actions that oppose Value 7 – Progress is made by way of finding solutions, not by finding personal fault:
- Being quick to assume that people "make defects" and searching for someone to blame when things go wrong.
- Being judgmental when faced with quality problems.
- Not creating the appropriate environment to allow or motivate employees to find solutions to problems.

Actions that oppose Value 8 – Everyone is involved:
- A mistaken belief some employees have that they are not part of "making quality happen".
- Managers not given specific quality improvement objectives.
- Sales-people (especially of major equipment) not held responsible for ensuring that the feasibility of achieving customer's requirements is adequately reviewed and agreed prior to order acceptance.
- Employees avoiding taking on a responsibility that causes change, particularly change in pursuance of quality improvement.

Actions that oppose Value 9 – Quality management is integrated with overall management:
- The belief that implementation of quality management is a project with a beginning and an end, rather than an ongoing process of company management.
- Assuming quality management to be a separate programme and for quality procedures to serve their own end such as ISO certification.
- Believing that effective quality management can be achieved through inspection for conformance.

- Not understanding that quality management is that part of the organisation's overall management system intended to ensure that it can meet customer requirements consistently through utilizing quality problem prevention, quality control and continuous improvement.

Actions that oppose Value 10 – People are the most important resource:
- Not continually investing in people by way of quality education, skills and competency training.
- A mistaken belief that quality is capital intensive and not people intensive, e.g., thinking that new expensive manufacturing equipment will alone result in good quality.
- Not giving recognition to individuals and teams who distinguish themselves in achievements and contributions that directly relate to key priorities and values set by the organisation.

Answers to Questions: Chapter 3

3-1

Among the important actions that managers and leaders can take to positively influence a company's quality management culture are the following:

(1) Adopt and mobilize the ten principle values of a quality management culture:

Top management can positively influence their company's quality management culture by personally adopting the ten principle values of a quality management culture in their actions, and by encouraging all employees to do the same in their daily routines. This can be practically achieved by mobilizing the values of a quality management culture in specific actions in preventing quality problems and in making progress in quality improvement. For example, problems can be prevented through an internal supplier actively identifying and fulfilling their internal customer's requirements. This mobilizes value 2: know your internal customers and their requirements, and support them.

(2) Stipulate the specific action required, focus the action with a performance metric and a target, and introduce frequent measurement and monitoring of the metric:

Managers and leaders need to do more than pronounce or stipulate the specific action that is required; they need to positively influence employees' actions and behaviour by focusing the action with a performance metric and a target, and must introduce frequent measurement and monitoring of the performance metric. The performance metric may be response time to answer customer queries, performance against agreed delivery schedule, or time to handle customer issues and complaints to satisfactory resolution, etc.

(3) Regularly monitor response and follow-up with appropriate actions or adjustments as necessary:

Response to the results of the action must be regularly monitored by managers and leaders, and follow-up actions or adjustments made as necessary because employees will very likely repeatedly relapse back into their old habits or behaviour before the new behaviour becomes engrained as the new way of doing things. Appropriate response by managers and leaders to the results will remind employees of the importance of needed results as well as encourage positive changes in the attitudes of employees, and reinforce or maintain the new focus of the employees.

(4) Appoint "quality champions":

The transformation to a quality management culture requires a lot of management in itself to maintain momentum of action, and therefore it is hugely helpful for top management to appoint senior-level "quality champions". These senior-level personnel comprehend quality management and have the necessary progressive attitude, approach, and knowledge of quality tools and techniques. They have enthusiasm and perseverance to overcome objections and obstacles.

The organisation's quality management culture must be enacted at all levels of the organisation therefore these "quality champions" must have the presence and ability to engage with employees at all levels, and have the know-how to remove organisational and personal barriers to progress.

3-2

Examples of how Visual Management in a manufacturing organisation is used to help control and achieve quality requirements and influence a desirable quality management culture:

In the workplace:

- Lines painted on the floor demarcate areas, and, accompanied by description boards, the use of specific areas is indicated, e.g., for product awaiting a certain process, and non-conforming product.
- On notice boards, visual displays are used to communicate quality measurement results to facilitate quality awareness and the need for

improvement. For instance: completed Run Charts are used to indicate trends, Histograms and Pareto Charts are used to highlight problems, and charts displaying the effect of certain actions taken to improve quality performance are used to facilitate quality thinking.

- Clear and vivid posters and notices are used to communicate information such as the organisation's mission statement, quality policy, strategic goals, critical issues, and important events.
- Notices and photographs are used to recognise involvement and effort and to acknowledge successes.
- Information (in hardcopy or softcopy), such as new or changed technical drawings and process procedures, are accompanied with colour coded sheets on which the recipient must provide feedback, e.g., yellow to indicate that clarification or further information is required, white to indicate that everything is understood and satisfactory.

At the workstation:
- Visual portrayal of instructions, visual aids (e.g., showing neatness of wiring) and visual quality standards (e.g., indicating aesthetic properties that are acceptable and not acceptable) are used to clearly communicate quality requirements. These visual means are useful for communicating difficult to describe quality requirements and characteristics. The visual means may be physical samples, sketches, simple drawings or photographs.
- Quality control activities are clearly visible through the use of measurement tools for checking process parameters, quality dimensions and characteristics. As needed, quality data capture forms, Check Sheets, Run Charts and Control Charts are at hand.
- Red/green lights are used to indicate go/no-go process control conditions, poka-yoke mistake-proofing devices are in use and keyed-fits are employed to avoid mistakes.
- Warning signs and key point charts are used to remind people to take certain quality actions.
- Quality status identification is in evidence by way of labels or colour coding to indicate acceptable and not-acceptable product status.

- Shadow-boards indicate where to place tools and measurement gauges, and what of these are missing.

3-3

Key point summary of the Human Performance System model:
- The performance of the work performer in a work process can be viewed in terms of a Human Performance System (HPS).
- The HPS is a model that describes the variables influencing the behaviour of a person in a work process.
- The HPS is governed by the behavioural law that people's behaviour is affected by consequences.
- The HPS has the following components:
  - Work performer
  - Input to the work performer
  - Output produced by the work performer with the given inputs
  - Consequences for output produced, and consequences for actions taken to make the output
  - Feedback given to the work performer of resulting consequences
- The consequences are interpreted by the work performer as either positive or negative; this interpretation is the key to understanding the work performer's future behaviour.
- The person's behaviour can be influenced by many things and it is advisable to assess each component of the HPS against an ideal HPS.

3-4

Conditions for each component of the HPS are described below for an ideal Human Performance System model:

Work performer:
- Has the skill and capability to do the job (mental, physical, and emotional)
- Knows where to get job-related information
- Possesses understanding of why the job is important

- Is willing to perform, given the incentives available

Input to the work performer:
- Performance expectations are clear
- Necessary resources for the work are in place
- Instructions are clear
- Received components do not need adjustment, rework or clean-up
- No or minimal interference from extraneous demands

Output produced by the work performer with the given inputs:
- Appropriate criteria and standards are known with which successful performance can be judged
- Output is aligned with business success
- Output performance requirements motivates possibility of further personal development

Consequences for output produced, and consequences for actions taken to make the output:
- Consequences and incentives encourage expected performance
- Consequences are fair, timely, and consistently applied
- Few, if any, negative consequences or disincentives to perform

Feedback given to the work performer of resulting consequences:
- Feedback is expressed in terms of job performance; it focuses on how well or how poorly the job is being performed
- Feedback is frequent and relevant
- Feedback is easy to understand and actionable

3-5

Powerful messages that could affect a quality management culture can be directly and also indirectly communicated by senior management in the manner in which they respond to quality issues, and in actions they take.

For example, senior management that are quick to respond constructively with an open mind to finding ways to improve quality, reflect interest in quality and quality improvement, and this interest radiates throughout the company to encourage a quality management culture.

Further encouragement is communicated from senior managers when they support teamwork, co-operation, openness and honesty in the management of quality, and when they insist that quality improvement stems from using facts to finding and eliminating the root cause of quality issues and problems; with this support and encouragement, employees are more likely to openly co-operate to advance quality and quality improvement.

Powerful positive messages can be indirectly communicated by senior management practicing "Management by Walking Around" (MBWA). Through MBWA the senior manager shows that he is interested and that he cares. The perception is created among employees that "the boss is interested and fully aware of what goes on"; this leads to a healthier and honest communication where the boss does not receive filtered information, or is told half-truths.

On the other hand, senior management that are slow to react or do not respond to quality issues reflect disinterest and this disinterest radiates throughout the organisation to detract from a healthy quality management culture.

Further detraction from a healthy quality management culture is communicated through senior management that repeatedly jumps to conclusions possibly influenced by personal prejudices knowing little or nothing of the actual facts, and that blame and penalize employees unfairly; this influences or drives employees to protect themselves, and employees are unlikely to openly co-operate (to advance quality and quality improvement) for fear of exposing themselves.

Answers to Questions: Chapter 4

4-1

The necessary conditions for an effective closed-loop quality control system are as follows:

1. The quality requirements of the output of the system must be known and absolutely clear.
2. The ownership of doing, observing/inspecting, and use of feedback to control quality is important; a single point of ownership enables harmony, co-ordination, and timing between the components of doing, observing/inspecting, and use of feedback to control.
3. When there is no single point of ownership of the components of doing and of observing/inspecting, and of use of feedback to control, it is imperative to ensure that there is excellent harmony, co-ordination, and timing between these components in order to maintain an effective closed-loop quality control system.

4-2

The basic understanding required for effective quality management system implementation is that actual implementation can only be done by the people in the organisation that do the work; these are the owners of the various work processes, and the quality management processes operate in these work processes. Top management plays a pivotal part in getting the functions of the organisation to accept ownership of quality management processes; their important actions are:

- Making clear to employees that they fully support every quality management process and are committed to making every quality management process successful.
- Ensure that education is given so that employees comprehended the intention of each quality management process, and ensure that guidance is given to ensure the correct outcome of each quality management process.

- They insist that quality management processes are applied in normal work routines, and take interest in their implementation.

4-3

Efforts of the Quality Management Department should be concentrated on customer satisfaction, identification of risks that could impede customer satisfaction, prevention of quality problems, and continuous improvement. Therefore a summary of the duties performed by a developed Quality Management Department are as follows:

- Ensuring that customer needs and requirements have been accurately identified and that the organisation is consistently working at meeting these needs and requirements.
- Ensuring that quality standards, processes, policies, checks and product inspection and test plans are in place.
- Organising regular (e.g., bi-weekly) internal quality audits to determine continual implementation effectiveness of quality management processes.
- Verifying that applicable technical standards are interpreted and applied correctly.
- Performing or arranging product validation and proving tests.
- Holding or participating in design reviews to ensure that relevant design milestone criteria have been reached.
- Identifying quality verifications and measurement, inspection and test tools.
- Providing or arranging training and education to enhance quality control at the workplace, and in the use of measurement techniques and Basic Quality Tools such as Check Sheets and Pareto Charts.
- Collecting and analysing data for identification of trends and opportunities for improvements, and encouraging others in the organisation to do the same.
- Playing a major role in quality problem solving and encouraging continuous improvement.
- Supporting the development and maintenance of customer focus within the organisation.

Answers to Questions: Chapter 5

$\boxed{\textbf{5-1}}$

The following emphasis in the responsibilities of top management is made clear in the latest revisions of ISO 9001 (GB/T 19001) quality management system requirements:

- Top management have responsibility for the leadership of the quality management system, and effective quality management requires their involvement and commitment.
- Top management must ensure that customer requirements are understood and met with the goal of improving customer satisfaction.
- The quality policy must identify the main goals of the quality management system and create a background for establishing quality objectives.
- Quality objectives, that support the quality policy, must be measurable and communicated throughout the organisation.
- Top management must set up an effective system of communication to ensure effective operation of the quality management system.
- Top management must regularly review the quality management system to make sure that the goals are being achieved, and to look for ways to improve its suitability, adequacy and effectiveness. The review must include assessing opportunities for improvement and the need for changes to the quality policy, quality objectives and quality management system.

$\boxed{\textbf{5-2}}$

Top management should ensure that an effective contract review process is in place for ascertaining that the organisation has an appropriate technical solution, and the capability of meeting customer requirements. Contract review process activities are briefly as follows:

- The contract review is performed at the appropriate time, i.e., upon receipt of an enquiry to supply or a request for quotation, and prior to the organisation submitting quotes and tenders, accepting contracts or orders, and accepting changes to contracts or orders.

- To ensure that the organisation can satisfactorily provide the required solution for the customer and meet the necessary technical requirements, a technical review is performed as part of the contract review process.
- Verification is conducted to ensure that needed resources are available, including appropriate equipment, facilities, skilled and competent manpower, and materials.
- A check of the ability to produce within the customer delivery time stipulated, taking into account such things as capacity to produce and supply-delivery times of bought-in items.

5-3

Key Performance Indicators (KPIs) can help in the achievement of strategic goals, objectives and performance improvement initiatives of an organisation, including those that are quality management related, in the following way:

- Strategic goals, objectives and performance improvement initiatives have critical success factors – performance objectives are determined that are linked to and directly support these critical success factors of the organisation.
- Each performance objective is assigned a (key) performance indicator. Usually between five and eight Key Performance Indicators are agreed with each of the organisation's management and senior personnel.
- Incentive is provided to motivate management and senior personnel to take action, and they are held accountable for the achievement of each KPI.
- The organisation's management and senior personnel apply themselves to achieve their performance targets.
- At a predefined time-period, as determined by the nature of the job and its impact on the organisations critical success factors,

performance against each KPI is reviewed. This is performed by the subordinate's manager by way of a formal "Personal Performance and Development Review".

- The review of the subordinate's achievements with respect to his performance targets is used as a motivational means as well as to draw out actions to improve or develop the performance of the subordinate; it is also used to identify and remove obstacles standing in the way of his performance – e.g., lack of knowledge, or resource issues.

- KPIs can be used to measure and track the degree to which monthly, quarterly, or annual performance goals and targets are being met. The examination for trends will also lead to the identification of areas showing high levels of good or bad (quality) performance, or areas requiring improvement. From these trends, decisions can be made to address areas that require attention.

5-4

Below are quality-related performance indicators specific to the following concerns of a business: customer satisfaction, profitability, product conformance, product design and development, equipment, personnel, quality management.

- Customer satisfaction:
  - Performance against promised delivery
  - Reduction of technical and commercial complaints
  - Achievement of stated product reliability
  - Reduction in warranty claims
- Profitability:
  - Reduction of the Price of Non-Conformance
- Product conformance:
  - Reduction of number of defects
  - Reduction of concessions raised
  - Reduction of rework and scrap
- Product design and development:
  - Design reviews completed and risks identified and removed

- Reduction in number of engineering changes
- Equipment:
  - Improvement of plant Overall Equipment Effectiveness (OEE)
- Personnel:
  - Employee skills and competency reviews performed
  - Training provided to improve skills and competencies (per skill gap identified)
- Quality management:
  - Number of corrective, preventive and improvement actions
  - Response and effectiveness closing identified corrective and preventive actions
  - Quality improvement projects successfully completed

5-5

Reasons for and examples of the typical type of internal communication that supports quality management:

1. Provide employees with the information they need to do their jobs effectively. E.g., drawings, procedures, process and work instructions, and quality checklists
2. Provide employees with clear standards and expectations for their work. E.g., specification tolerances on technical drawings; process control limits; performance targets; visual and non-visual quality standards
3. Ensure employees are informed about anything that concerns them. E.g., management briefs and information notices concerning quality achievement and non-achievement; quality performance measurement information; electronic workflows for contract review and customer complaint handling.
4. Give employees feedback on their own performance. E.g., Personal Performance and Development Reviews, and Skills and Competency Reviews
5. Provide employees emotional support for difficult work. E.g., supportive communications that ranges from giving encouragement to counselling (alleviating work related stress and

frustrations, and addressing consequent problems causing quality errors arising from distraction of focus)

6. Allow employees to understand the situation in their work area and the organisation. E.g., Pareto Charts, Histograms, and Defect Concentration Diagrams showing occurrence and prominence of defects; quality key point charts drawing attention to issues of importance

7. Help employees maintain a shared vision and a sense of ownership and pride. E.g., discussions, notices, awards (reminding employees of the company vision and quality goals, sharing customer feedback on quality, and giving employees quality excellence awards)

5-6

(a) Typical areas in an organisation from where information can be collected to facilitate quality improvement are: sales and marketing, customer service, design engineering, production, quality control, logistics, installation, accounts.

(b) Management can encourage the collection and use of information to make advancements in quality improvement by:

- Making sure that employees, especially staff and production workers, are given the appropriate instructions and means to collect information.
- Organising the education of employees in the use of Basic Quality Tools.
- Being aware whether employees resist reporting quality information because of a blame and penalty culture, and making changes to remove causes that make employees suspicious that reported quality information could be used against them.
- Working with the employees to develop measurements that have meaning for both employee and the company through, e.g., Key Performance Indicators and performing employee performance appraisals.

Answers to Questions: Chapter 6

6-1

Quality Costs are the costs associated with ensuring and assuring quality, and the loss incurred when quality is not achieved. Quality Costs comprise Prevention, Appraisal, Internal and External Failure Costs.

- Prevention Costs: The costs of all activities aimed at preventing, avoiding or minimizing poor quality in products and services.
- Appraisal Costs: The costs incurred before delivery or handover to customers associated with measuring, evaluating, or auditing products or services to assure conformance to quality standards and conformance requirements.
- Internal Failure Costs: The costs incurred before delivery or handover to customers resulting from products or services not conforming to requirements or customer needs.
- External Failure Costs: The costs incurred after delivery or handover to customers resulting from products or services not conforming to requirements or customer needs.

6-2

Typical activities and practices that contribute to failure costs but exclude scrap, rework, customer complaints and repair, and are commonly considered to be normal operation costs, are as follows:
- extra operations (e.g., touch-up, de-burring)
- downgrading
- expediting
- downtime caused by quality problems
- time with dissatisfied customers
- lost sales arising from reputation for poor product

6-3

Some of the prevention activities and practices that would bring benefit to an organisation:
- new product design reviews
- process capability studies
- internal quality auditing (to prevent potential quality problems from occurring)
- data gathering, analysis, reporting and use
- quality improvement projects
- Statistical Process Control
- quality education and training

Answers to Questions: Chapter 7

7-1

The need for Quality Planning in an organisation could be brought about by the following reasons:
- due to the necessity to develop the quality management system and its processes;
- from issues identified in the company's strategic planning process or management review process such as company organisational change, new quality objectives or change in emphasis of the quality management system;
- as a result of customer order special requirements, and product containing special or non-standard requirements;
- from analysis of data where trends are indicated and opportunity for improvement and quality improvement projects are identified;
- due to repetitive or recurring quality problems (causing "fire-fighting");
- as a result of weaknesses found during supplier assessments;
- due to special requirements applicable to a bought-in item;
- from internal quality audit findings where, for example, risk due to major quality management shortcomings or the absence of controls are identified.

7-2

The main purposes provided by Design Reviews are as follows:
- To determine if the product will work as desired and meet the customer's functional and performance requirements, as well as regulatory and statutory requirements, including those of safety.
- To ensure that satisfactory reliability is achieved and improved where possible.

- To ensure that the build-up of tolerances is taken into account in assemblies and components, obviating the need for costly customization to achieve the necessary fit.
- To standardise on components and assemblies as much as possible to avoid the myriad of costly variations and the impact on drawing control, tooling, jigs and fixtures, measurement and test.
- To determine if a new design is realistically economically producible, bearing in mind the resources available, e.g., the capability of the factory's equipment and the competency of available manpower.
- To ensure that information derived from previous designs, and feedback from actual use by the customer, are taken into account to develop better designs.
- To determine if the new design is maintainable and repairable, and that replacement assemblies and components are interchangeable.

7-3

Shortcomings and risks inherent in relying on end-of-process inspections by an independent inspection function are as follows:
- End-of-process or final inspections can incur high Appraisal Costs and fail to prevent internal failures or their recurrence.
- Failure to prevent internal failures leads to Internal Failure Costs.
- End-of-process inspection fails to screen out all internal failures* therefore the risk of external failures and External Failure Costs will increase.

  *Note that even 100% inspection at the very end of production has a probability of not detecting some defects; this is because defective items will slip through unless the inspection process is itself 100% effective, and this will, by the nature of human and measurement errors, have a probability of uncertainty.
- When parts are found defective at the very end of production it is often difficult to pinpoint the specific cause of defects. In addition, when defects are found by an independent inspection function long after they were caused, it usually is difficult to ascertain accurate corrective action that addresses the root cause of the problem.

- Once components and fabricated items are processed, i.e., they are machined, heat treated, welded or painted, they have "added value". Because of this, if found defective, they cannot be readily scrapped and often their rework or re-processing is very costly. Their fate will need to be carefully considered. This typically results in a protracted disposition cycle to determine what to do with the non-conforming work, and the event diverts personnel from productive work and results in a build-up of work-in-progress.

7-4

The general steps concerning the collection and analysis of data for use as quality information, are as follows:

1. Decide what data would be relevant for use as quality information:
   Relevant data would be that which is used to ensure customer satisfaction and quality management system effectiveness; it is used to guide and improve human performance, to improve product design, product conformance and customer service, and it is used to identify risk and where efforts need to be placed to reduce PONC.

2. Identify collection points for this data:
   Information must be collected from many areas of the organisation, e.g., from sales and marketing, customer service, design, production, logistics, field service, accounts, quality control and human resource.

3. Design a data collection method:
   Methods for collecting data must suit the situation. Favoured methods for collecting data during production are, for example, defect tally sheets, check sheets and non-conformance reports, and, for capturing customer complaints and response times, customer complaint handling registers and service failure investigation reports.

4. Arrange for the data to be captured and communicated:
   The capturing of data must be planned, and a routine must be established for communicating this data for use and analysis.

5. Conduct timely analysis to reach intelligent fact-based decisions:

Quality tools are used to aid analysis, problem identification and resolution, and quality improvement. Scatter Diagrams, Control Charts and Flow Charts, Pareto Charts, and Cause and Effect Diagrams are effective for problem analysis.

7-5

The follow are some reasons why quality education is important for top and middle management:

- The management of quality has a huge impact on customer satisfaction and the development and the profitability of the organisation, therefore the comprehension of quality management by top management is vitally important.
- Top management needs to understand the language of quality and the concepts of quality, what employees need to do to achieve the organisation's quality standards and objectives, and how quality affects goals of customer satisfaction and ongoing success of the organisation.
- The quality education will help senior people understand their role in developing a prevention-orientated quality management system, how they are supposed to react to non-conformances, and how they can encourage improvement in the quality process.
- Middle management bears a great responsibility for maintaining and developing quality management because this group of managers represent and interpret the established management policy. They are the key persons in communicating and tracking the various kinds of goals and in making information flow up and down, and need to show their quality commitment to their people every day.
- Top management in organisations certified to ISO 9001 (GB/T 19001), are held as overall responsible for quality management.

7-6

The purposes and aims of internal quality auditing are as follows:

- Internal quality auditing provides a means to verify the effectiveness of application and control of the quality management system processes, and to assess process and system strengths and weaknesses, and therefore to identify actions needed to improve process and system effectiveness.

- An effective internal quality audit process will uncover risks associated with the lack of control, and verify that corrective and improvement actions are effective and performed in a timely manner.
- Internal quality audits will enable management to obtain a measure of its own effectiveness in controlling the operation of the organisation in the manner intended.
- Internal quality audits, as a requirement of ISO 9001 (GB/T 19001) are necessary to determine whether the organisation's quality management system;
    - conforms to the requirements of the International Quality Management Standard,
    - conforms to the quality management requirements established by the organisation,
    - is effectively implemented and maintained.

7-7

The following serves to outline what needs to be considered in an effective internal quality audit program:

1. An "Internal Audit Schedule" is developed at the beginning of each year indicating what will be audited and during which week in the year each targeted process will be audited.
2. The internal audit process is an ongoing activity, continually auditing different parts of the system and covering all areas of the organisation. There could be two or more of these audits of different processes of the organisation spread out over a month.
3. Audits will be highly focused and of short duration; the short duration audits are easier to manage and cause minimal disruption to operations.
4. Areas where the audit findings are adverse, receive more frequent

audits until such time as the audit results stabilise; particularly important activities are also audited more frequently.

5. The selection of what and when to audit is based upon status and importance.

   - Status concerns how a particular department or process is performing against expectations and established goals and objectives. The status is reflected in, for instance, the department's or process's quality performance history, changes in process or equipment or key personnel, and re-structuring or re-organisation of the department.

   - The essence of the meaning of importance is with regard to the power to produce an effect, e.g., the importance of the inspection process lies in its effectiveness in preventing non-conforming products from being dispatched to the customer.

6. An effective "Internal Audit Schedule" does not only contain audits with a focus on obvious shortcomings or weaknesses in an organisation, it also contain audits focused on the areas that could potentially be a root cause for non-conforming or risk situations, as well as areas critical for maintaining product conformity.

7. The "Internal Audit Schedule" is a living document that can be revised throughout the year as may be required to take into account changes and findings from such things as management review outputs, customer concerns, internal quality issues, supplier and sub-contractor quality problems and organisational changes.

8. Many of the processes identified as needing to be audited span different functions or departments, and their effectiveness can only be determined when all things that shape and support the overall process are taken into account. The audit for process effectiveness must include interface activities with other processes and take into consideration the requirements of internal suppliers and internal customers, i.e., the quality audit must take a process-based approach.

Answers to Questions: Chapter 8

8-1

- Processes are most effectively arranged to satisfy customers' needs and create value through an arrangement of "core" processes supported by a competent set of "support" processes.
  - "Core", also called "business" processes, are the value creators;
  - "Support" processes are management and service processes that help the "core" processes to create value.
  - Quality management processes are an intrinsic part of both "core" and "support" processes.
- The output of each "core" process serves as an input to the next "core" process. The downstream process is the internal supplier of the upstream process. The upstream process is the internal customer of the downstream process.
- Internal and external suppliers, including those providing "support", provide inputs according to the defined input requirements of their customers; these may be internal customers or end-customers.
- When the supplier knows the requirements of its customer, it is able to identify and control the delivery of its own input requirements to best achieve its defined output to meet the requirements of its customer.

8-2

Key activities that enable, support and improve operational performance are:
- Ensuring that each process in the process-chain has its output requirements clearly defined, and for each process output, the supplying process must carry out checks to ensure that its internal customer requirements are met.

- Analysing and optimally balancing the configuration of production lines for the most practically efficient flow that avoids build-up of work-in-progress and excessive handling of material and product.
- Optimising the layout of each workstation with due regard to the movement of material and the operator, to create order and a safe environment, to minimize movement and to enable quality control actions (inspection, identification, recording, segregation).
- Ensuring competent performance of workers through their continual evaluation, training and education. Training is given in on-the-job skills, and quality control actions specific to the work. Education includes quality awareness, personal health and safety in the workplace, environmental practices and application of quality tools, as required
- Setting up a Local Area Network in work areas so that operators can easily access current, correct, approved and controlled information required for their work, such as process procedures, technical information, and quality control instructions.
- Establishing a system to track work progress on computer systems through all phases of manufacturing and quality control using an information system with Material Resource Planning functionality.
- Organising help-call buzzers at workstations for operators to use to alert the supervisor of, for example, material and machine problems, and to alert the quality control technician of quality issues.
- Establishing measurement and information collection systems to report types and number of defects, to indicate process variability, and to report the overall utilization of facilities, and use this information to promote and advance continuous improvement.

8-3

Advantages of piece-work schemes:
- Encourages labour output to meet and exceed piece-work targets.

Disadvantages of piece-work schemes:
- The worker's main concern is about quantity and not the quality. The operator, being paid according to the number of parts he

produces, would rather allow a defective part to pass than to penalize himself.

- Introducing quality grading in an attempt to overcome the negative effect of the operator penalizing himself when declaring non-conforming parts does not encourage good quality.
- The piece-work scheme requires parts produced to be examined by another party; when the other party identifies non-conformances that results in loss to the operator, friction is fuelled on the production floor.
- The task of determining piece-work targets encourages gamesmanship; when the operator sees the industrial engineer timing him, the operator could reason that it makes more sense to conceal how much he could do, so he may deliberately slow down his work rate.
- Piece-work schemes discourage suggestions from operators to improve productivity.
- Piece-work schemes apply to individuals; this discourages individuals from spending time in worker-to-worker participation for better quality.
- It is difficult to measure a person's overall value-adding benefit based solely on the units they can produce in a given period of time.
- Piece-work schemes require costly administration, difficult to justify.

8-4

PDCA is an improvement method and a way of thinking. PDCA involves four phases – Plan, Do, Check and Act.

- Plan: Firstly, the problem must be clearly understood therefore the current situation needs to be examined and data collected and analysed. Once the problem is understood, what needs to be done be determined. An experiment or test method then needs to be determined. In the process of formulating what needs to be done, to force more careful up-front thinking, the predicted results should be written down; this also helps to set better experiments and helps with the learning process.

- Do: There may be many things to do that emerge from the "Plan" phase, but, in order to avoid confusion by doing too many things at once, prioritize what to do. Try the new method on a small scale.
- Check: Evaluate the new method to determine whether it has resulted in the predicted or expected change, and summarize what was learnt.
- Act: If the experiment produced an effect in line with the aim or objective, implement and standardise the new method. If the results from the "Check" phase were unsuccessful, make a new plan.

8-5

A manufacturing organisation has refurbished an automated welding machine and wants to know whether refurbishment has improved OEE.

Summary of given information:	Before	After
Planned operational time (hrs) out of 24 hrs	18 (1080 mins)	18 (1080 mins)
Cycle time	$\frac{1}{15}$ min or 4 secs	$\frac{1}{12}$ min or 5 secs
Actual operating time	12 hrs 10 mins (730 mins)	15 hrs 20 mins (920 mins)
Actual welds made	9860	10190
Welds passing quality control tests	9120	10156
Note: Potential weld-rate is given as butt-welds per minute, but the reciprocal cycle time per butt-weld is required in the equation.		

Calculations:

Equations	Before	After
Availability % : $\dfrac{\text{actual operating time}}{\text{planned operating time}} \times 100$	$\dfrac{730}{1080} \times 100$ = 67.6%	$\dfrac{920}{1080} \times 100$ = 85.2%

Performance % : $\dfrac{\text{cycle time x total units started}}{\text{actual operating time}} \times 100$	$\dfrac{9860}{15 \times 730} \times 100$ = 90.0%	$\dfrac{10190}{12 \times 920} \times 100$ = 92.3%
Quality % : $\dfrac{\text{good units produced}}{\text{total units started}} \times 100$	$\dfrac{9120}{9860} \times 100$ = 92.5%	$\dfrac{10156}{10190} \times 100$ = 99.7%
OEE % : Availability % × Performance % × Quality %	67.6 × 90.0 × 92.5 = 56.3%	85.2 × 92.3 × 99.7 = 78.4%

Discussion of results and suggestion of ensuing improvement actions:

- The refurbishment resulted in significant improvement in Availability from 67.6% to 85.2%.
- Machine Performance increased a little from 90.0% to 92.3% but could have been better had the refurbishment not slowed down the number of butt-welds that the machine could produce per minute (decreased from 15 to 12 butt-welds per minute).
- Quality improved satisfactorily from 92.5% to 99.7%.
- Further improvements in the OEE of 78.4% may be attained through the following:
  - Record all significant stoppage events – provide the operator a 24 hour log sheet on which to record machine downtime events; analyse and take action to minimize or eliminate these events.
  - Perform tests to determine whether the Performance of the machine can be improved through incremental increases in butt-welds per minute. (Use the PDCA method.)

8-6

Discussion of the following causes of the cost of poor quality:

(a) Non-conforming product found during final inspection or testing

(b) Past due date receivables

(c) Expediting

(a) Non-conforming products found during final inspection or testing become of great concern because they have usually undergone much processing and therefore their value can be very high. Because of this, the disposition decision needs to be carefully weighed. For instance, rework would be difficult and complicated where gearboxes are found excessively noisy or running too hot during final testing; scrapping is very costly of a fully machined housing having incorrectly positioned mating holes – it may be less costly to filler-weld, dress, and re-drill the holes. The process of deciding what to do and the ensuing actions attract time and costs, and also increase the amount of revenue held in working assets.

(b) Payments on Accounts Receivable for products and services that are not made within a specified time result in "past due date receivables". Poor quality control can cause many invoices not to be paid, or not to be paid in full. Some of the common reasons for delayed payment due to poor quality control include:

- A dispute regarding product quality conformance or quality of work performed.
- An error in invoicing, e.g., incorrect particulars or an incomplete or inaccurate statement.
- Incomplete delivery, e.g., incorrect number of items or a missing technical test report.
- Use of an incorrect invoice form.

(c) "Expediting" is incurred in activities taken to hurry the attainment of a required outcome. Almost all expediting expenses are as a result of failure to control quality. For example, the delivery of a product was promised for a certain date and, due to not considering all necessary product realisation requirements, or due to the unforeseen re-processing of defective product, more time than planned is taken to produce the product. In an attempt to recover this time, overtime is incurred, and/or express transport is employed; these are typical expediting expenses that can be avoided by better quality management.

8-7

In order to promote "Lean Thinking", it is helpful to firstly understand that the following broad principles guide "Lean Thinking":

(a) The expenditure of resources for any goal other than the creation of value for the end customer is regarded as wasteful and thus targeted for elimination.

(b) The implementation of smooth flow exposes constraints and quality problems, and thus action is directed towards overcoming these issues.

Secondly, it is helpful to understand that the "7 Wastes" of Lean Manufacturing need to be targeted for elimination; these are:

1. Over-processing / over-complication
2. Idle time / waiting time
3. Bad quality or defects
4. Overproduction
5. Excessive inventory
6. Unnecessary handling, movement and transport
7. Unnecessary motion

8-8

The following are quality-related benefits an organisation can secure from implementing the 5S/5C practices:

Sort / Clear-out:

- Removal of all unnecessary items from the workplace results in less clutter and orderly thinking; needed items, e.g., jigs and measurement gauges, become readily accessible.
- People working in the area are sensitive to non-moving components stacking up such as work awaiting quality disposition or further processing; this tends to spur the taking of action.

Set in order / Configure:

- Locating needed things in clearly identified places in the work-area means that tools, jigs and measurement gauges are quickly available when needed.

- The opportunity for errors and quality-related risks are minimized when, for example;
  - there is place to keep correct and controlled drawings, process procedures, work instructions, and quality inspection instruments;
  - there are identified locations to store necessary tools and measurement gauges such as shadow boards and cabinets;
  - there are clearly identified areas in which to place components awaiting inspection or found non-conforming and awaiting quality disposition
  - there is consistent content of documentation accompanying each job

Shine / Clean and Check:

- Cleaning and checking the work area, and everything in the work-area, brings the benefits of early identification of any problems as well as concerns and abnormal conditions that may impact on quality (and on general work-area performance).
- The practice of making sure that the status of material is identified, or that the material is placed in a clearly identified location, prevents the inadvertent use of non-conforming material.

Standardise / Conformity:

- Prevention of problems and waste is facilitated through people in the work-area routinely ensuring that the correct methods, standards, and checks are employed, and the right tools are available.

Sustain / Custom Practice:

- Ongoing support, encouragement and recognition from management bring the benefits of ensuring that quality gains obtained are sustained and further enhanced.
- Ongoing attention to job details, clear work instructions, and the training and education in necessary work-skills, quality and safety, enhances effectiveness of the performance of people in the workplace.
- The custom of maintaining tidy and organised work-areas boosts the "quality" image of the organisation.

Answers to Questions: Chapter 9

9-1

The purchaser's supplier selection team needs to determine that the potential supplier has the necessary capabilities to do the required work, to consistently meet quality standards and to control quality, and to deliver according to the required schedule and agreed cost. The team also needs to determine the supplier's limitations or possible weaknesses that could present possible risks in meeting the purchaser's technical, quality, delivery, and cost requirements. Therefore, and particularly concerning the potential supplier of complex and high value items for a major contract, the purchaser's team needs to be on the lookout for the following:

- Capability and adequacy of production equipment and facilities.
- Production capacity.
- Financial position.
- Operational staff technical knowledge and ability to control operations.
- Suitability and adequacy of inspection and test methods.
- Expertise of the supplier's employees regarding providing solutions to production problems.
- Attitudes of the supplier's employees towards their work and control of the quality.
- Care and maintenance of production equipment (relates to quality consciousness).
- Adequacy of the supplier's quality management processes and practices for controlling, e.g., technical documents, non-conforming material, purchaser's supplied tooling.
- Orderliness and cleanliness of the supplier's production premises (gives an indication of management's planning and control).
- Attitude towards safety and avoidance of disaster (relates to continuity of supply).
- Commitment of the supplier's management to quality.

9-2

Quality control and quality assurance conditions that the purchaser can stipulate in the "Contract to Supply" are such as:

- That evidence of special process controls must be supplied with each delivery, e.g., hardness tested samples after heat treatment, with proof of Brinell test indentations.
- That test certificates for safety critical goods (e.g., lifting chain) must be supplied with each delivery.
- That certain work such as welding and Non-Destructive Testing must be performed by certificated personnel.
- That the purchaser's Quality Assurance Representative (QAR) will perform "Surveillance Visits" to monitor the supplier's production processes.
- That release from the supplier's premises can only be performed by the purchaser's QAR after the QAR is satisfied that the goods conform to the purchaser's technical and quality requirements.
- That a Quality Plan must be followed during activities within the scope of the supply agreement. This plan could call for the application of specific controls, specific process work instructions, use of reference samples and workmanship standards, special procedures, specifications and standards, and specific stage-related inspections and tests. "Hold" and "witness" points could be identified where the supplier is required to arrange for the purchaser, or his representative, to approve or observe a certain activity. The Quality Plan could also stipulate requirements for "record of proof" that the supplier must retain as evidence that certain verifications have been carried out.

9-3

- A general reason for the purchaser to have a Supplier Development Programme would be the need for the purchaser to ensure that its suppliers develop the desired characteristics and abilities to provide products and services that meet all of the purchaser's requirements.

- The supplier could have all the necessary equipment and facilities but the work required to be done by the purchaser may demand the use of special purchaser processes that need to be taught to the supplier by the purchaser's experts.
- The purchaser could also be driven by circumstance or the lack of choice to source product from certain suppliers that have the basic capability to produce what the purchaser requires, but lack the expertise to do the work to the specific and exacting requirements of the purchaser.

9-4

- The assistance that can be given by the purchaser to a supplier is not through providing additional manpower to do the work, but by way of providing skilled manpower usually on an as-required part-time basis. This skilled manpower brings to the supplier expertise, knowledge and guidance in practices, processes and techniques where necessary.
- The assistance can be general and/or specific, and will depend on the development status of the supplier.
  - General assistance could be provided by way of guidance in workmanship standards, practices to eliminate the 7 Wastes, the use of basic quality tools, visual displays for measurement information, the development of a supplier development scoreboard, and visual controls for making problems, abnormalities, or deviations from standards visible to everyone.
  - Specific assistance could be provided such as in guiding the application of technical process specifications, or in explaining the practical implementation of requirements contained in national and international standards, and also in the guidance and training in the application of contract-specific processes.

Answers to Questions: Chapter 10

> ### 10-1

"Common" or natural causes of variation are inherent in a process; examples are an oven's thermostat regulating oven temperature, variation of raw material properties and machine wear.

"Special" causes of variation, also referred to as assignable causes of variation, are process disturbance events; examples are opening the oven door during heat treatment causing oven temperature to drop, incorrect raw material and machine needing adjustment.

> ### 10-2

All processes have inherent or natural variation, if this falls within a given specification or product tolerance the process is said to be capable. The study of process performance over a period of time will reveal whether a process is capable of producing an output that is 100% in conformance to the given specification or product tolerance. For this study, a "Run Chart" is useful to determine whether the process is in-control and centred, i.e., whether its output is consistent and stable, and whether the process values are generally centred between the specification's limits.

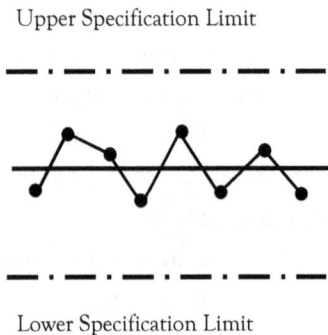

Upper Specification Limit

Upper Specification Limit

Lower Specification Limit

Lower Specification Limit

Run Chart 1:
Process centred and stable

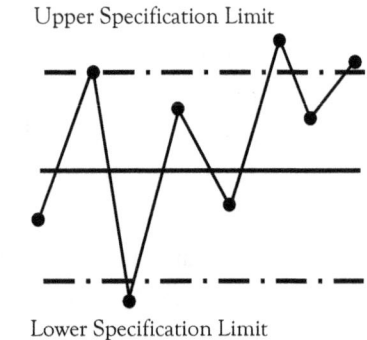

Run Chart 2:
Process not centred and not stable

Run Chart 1 illustrates a centred and stable process that can be deemed a "capable process".

Run Chart 2 illustrates a process that is not capable of producing an output that is 100% in conformance to the given specification.

10-3

With "common" and "special" causes of variation in mind, process improvement should be approached in the following manner:

A study of the process and interpretation of its variation is necessary. "Common" causes of variation are inherent in a process and are therefore always present, but the observed performance (process output) could include "special" causes of variation, i.e., process disturbance events. The "special" causes of variation must first be identified and eliminated.

A process could be stable and free from "special" causes of variation but produce unsatisfactory results with regard to a desired target or specification. Adjusting the process mean could bring it on target but if the "common" causes of variation produce an excessive variation relative to the desired specification, the process system inputs and influencing factors (man, material, method, environment, etc.) need to be studied with a view to improvement.

10-4

To determine whether a process has the capability of operating within an engineering tolerance, the process performance measurements of a stable and in-control process are compared with specification limits.

In order to do this, it is necessary to obtain a measure of the performance of the process, i.e., to determine the extent and nature of its variation. The process variation must be checked to be free from "special" causes of variation, and confirmed to be stable and predictable over a sufficiently long period of time to ensure that time-of-day/week/month or time-of-year effects are not biasing the measurements. The Averages and Range Chart is very effective for checking that the process is stable and in-control, and for detecting any "special" causes of variation.

When the process has been confirmed as being stable and in-control, process values (sample measurements) are plotted on a Histogram. These process values must accurately represent the process therefore they need to be randomly selected and sufficient in number. The amount of sample measurements taken can be determined by checking when the cumulative average of sample measurements flattens out to a steady cumulative average for a period of time.

The specification limits are entered onto the Histogram in order for a comparison of the spread of plotted process values to be made with the specification limits.

10-5

The organisation can ensure that their products and equipment have the capability to satisfactorily perform the intended application through the organisation's specifier or designer specifying with due regard and understanding to the full extent of the application of the products and equipment, and the environment, and the conditions under which they must operate. These can vary appreciably from one customer's site to another, therefore site-specific information concerning application, environment and operating conditions must be obtained, and the following must be considered:

1. The actual performance variation of the product or equipment:
   - The initial properties of the manufactured product would have natural variation affected by the properties of its constituent materials and components and the variations of the manufacturing processes. This variation should be controlled during manufacturing and verified for conformance to specification.
   - When various products or equipment-set building blocks, such as motors, gearboxes, couplers, drives, and controllers are brought together in an equipment-set, the capability of the individual

products in operation could be affected, and the use of similar but different constituent building-blocks would obviously cause a variation in equipment performance. These effects and variations must be carefully considered.

2. The variation in operational stress or load, and operating conditions:
   - The full nature of the stresses that would be encountered by the product or equipment during its operation in the field must be ascertained. "Nominal" quantities or values given by the customer for load, stress, electrical current and voltage, etc., are not actual or exact and could be appreciably different from the actual value or quantity, so the actual quantity or value and its spread or variation during operation must be established.
   - Operating conditions pertaining to the environment such as temperature extremes and rate of change, humidity, corrosion, vibration, shock, etc., and operating conditions pertaining to the man-controllable factors such as numbers of cold starts and aggressiveness of control by the work-force, need to be taken into account. The designer or specifier needs to take all "common" causes of applied stress or load into account including some "special" causes of stress or load that could arise such as due to momentary shock-loading.

3. A design safety margin:
   - A design safety margin should allow a reasonable margin between the product or equipment performance capability and the stress or load encountered by the product and equipment during operation.
   - Where it is known that the product strength decreases or decays over time, the margin between initial product strength and applied stress or load should be increased to prevent failure of the product within the guarantee period.

4. Equipment set-up and maintenance routines:
   The capability of the product or equipment in operation could be affected by set-up and maintenance routines. For instance, drive-belt or conveyor chain must be correctly pre-tensioned to enable it to provide specified performance.

Answers to Questions: Chapter 11

11-1

Uses of quality tools are indicated with a cross in the table:

	Understand processes	Problem identification	Problem visualization	Problem analysis
Flow Chart	X	X		
Check Sheet		X	X	
Histogram		X	X	X
Pareto Chart		X	X	X
Run Chart		X	X	X
Control Chart	X	X	X	X
Cause and Effect Diagram		X		X

11-2

(a) Purpose and operation of a Control Chart:

A Control Chart is used to communicate how a quality attribute or variable, or process performance changes usually over time, and to draw attention to when actions are needed to improve quality or reduce process variation.

A Control Chart displays statistically determined upper and lower control limits (UCL and LCL) drawn on either side of a process average or mean, and shows whether the collected process values are within UCL and LCL. When a process is operating in a stable and predictable state, process variation occurs between UCL and LCL; this variation is said to arise from "common" causes inherent to the process. When the process variation moves beyond UCL and LCL to an out-of-control state, this is due to assignable causes, also referred to as "special" causes.

(b) Averages and Range Control Chart ($\overline{\overline{X}} - R$ Control Chart):

The $\overline{\overline{X}} - R$ Control Chart comprises of two charts drawn together and used where sensitivity to process variation is an important quality control criterion. The statistically derived control limits provide a greatly improved probability of detecting out-of-limit measurements.

The Averages Chart focuses attention on the process average and is good for detecting out-of-limit situations, and the Range Chart is specifically designed for detecting changes in process variability. Process variation on the Range Chart beyond UCL indicates the process is not in statistical control, regardless of what the Averages Chart indicates.

11-3

- Calculation of $\overline{X}$, R, $\overline{\overline{X}}$ and $\overline{R}$ for the construction of an $\overline{X} - R$ Control Chart from the data given in question 11-3:

Subgroup	Process values taken every hour					$\overline{X}$	R
1	89.2	89.4	89.2	89.5	89.2	89.30	0.3
2	89.5	89.2	89.4	89.5	89.1	89.34	0.4
3	89.4	89.6	89.5	89.3	89.3	89.42	0.3
4	89.3	89.2	89.5	89.4	89.1	89.30	0.4
5	89.5	89.3	89.4	89.7	89.3	89.44	0.4
6	89.3	89.7	89.4	89.2	89.2	89.36	0.5
7	89.2	89.5	89.4	89.6	89.7	89.48	0.5
8	89.2	89.6	89.6	89.3	89.1	89.36	0.5
9	89.6	89.3	89.6	89.2	89.8	89.50	0.6
10	89.6	89.7	89.4	89.1	89.1	89.38	0.6
11	89.2	89.9	89.6	89.3	89.4	89.48	0.7
12	89.8	89.4	89.3	89.2	89.1	89.36	0.7
13	89.5	89.6	89.0	89.4	89.7	89.44	0.7
14	89.5	89.1	89.9	89.4	89.6	89.50	0.8
15	89.2	89.9	89.5	89.3	89.1	89.40	0.8
						$\overline{\overline{X}}$= 89.40	$\overline{R}$ = 0.547

- Calculation of UCL and LCL for $\overline{X}$ Chart, UCL and LCL for R Chart:

  See table 11-1 in Chapter 11 for $\overline{X} - R$ Control Chart Constants A_2, D_3 **and** D_4 for sample or sub-group (n) sizes of 5

  $\overline{X}$ Chart UCL = $\overline{\overline{X}} + (A_2 \times \overline{R})$ = 89.40 + (0.577 × 0.547) = 89.72

  $\overline{X}$ Chart LCL = $\overline{\overline{X}} - (A_2 \times \overline{R})$ = 89.40 – (0.577 × 0.547) = 89.08

  R Chart UCL = $\overline{R} \times D_4$ = 0.547 × 2.114 = 1.16

  R Chart LCL = $\overline{R} \times D_3$ = 0.547 × 0 = 0

- Construction of $\overline{X}$ and R Control Charts:

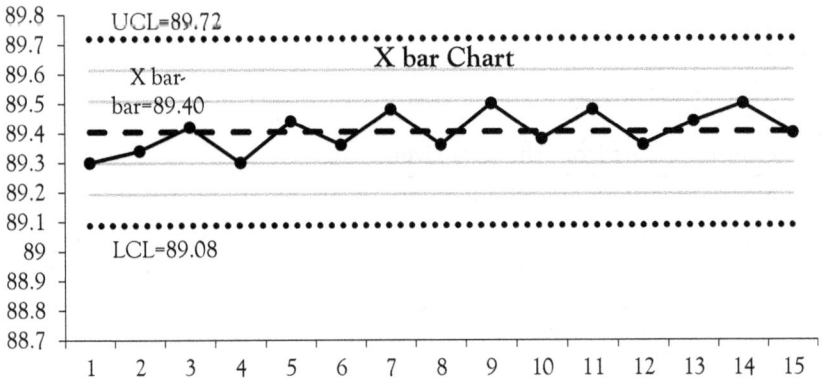

X bar Chart

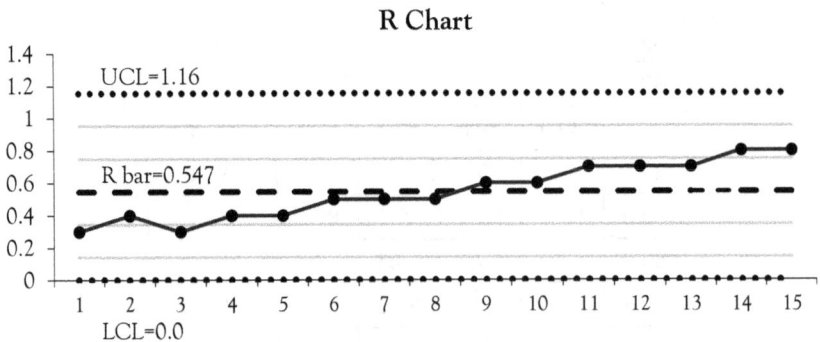

R Chart

11-4

(a) Determination of whether the machine with the performance results in answer 11-3 can maintain an engineering tolerance of 89.5 ± 0.7mm:

Tolerance that can be maintained = m ± $3\sigma_n$
  where tolerance midpoint = m and estimated standard deviation = σ_n
  m given as 89.5mm
  σ_n = $\overline{R}$ × applicable sub-group constant
  The applicable sub-group constant = 0.43 for the sub-group size of 5 used (from table 11-2 in Chapter 11, for n = 5)
  From Answer 11-3, $\overline{R}$ = 0.547
  Therefore σ_n = 0.547 × 0.43 = 0.2352
Tolerance that can be maintained = m ± $3\sigma_n$ = 89.5 ± (3 × 0.2352)
                                  = 89.5 ± 0.706

(b) Although the calculation indicates that the machine could just maintain the given tolerance, the machine variability is noted to steadily increase over time and this would be a major cause for concern because if no action is taken out-of-tolerance product will be produced, and produced in a relatively short period of machine operation. It is therefore advisable to investigate the reason for this steady degradation in the machine's precision capability. This may be due to an increase in machine running temperature caused by lack of lubrication, or perhaps due to wear-out of machine bearings. After taking the necessary corrective action to address the degradation, the machine process average should be reset from 89.4 to the specification mid-point of 89.5, and further monitoring performed to confirm the machine's stability.

A truer estimation of the tolerance that the machine can maintain will be obtained when the variability of the nature presented (i.e., the steady increase in variability over time) is removed or can be definitively controlled through following a strict procedure.

APPENDIX 1: TERMS, STANDARDS & CONCEPTS

Look-up List

1. Best practice
2. Blame culture
3. China Quality Award
4. Continuous improvement and continual improvement
5. Critical-To-Quality (CTQ)
6. Effectiveness
7. Efficiency
8. 5 Whys
9. Function
10. ISO 9000 family of standards:
    - ISO 9000:2015 (GB/T 19000: 2016), Quality Management Systems – Fundamentals and Vocabulary
    - ISO 9001:2015 (GB/T 19001: 2016), Quality Management Systems – Requirements
    - ISO 9004:2009 (GB/T 19004: 2011), Managing for the Sustained Success of an Organisation - A Quality Management Approach
11. ISO 19011:2011 (GB/T 19011: 2013), Guidelines for Auditing Management Systems
12. KAIZEN
13. Key Performance Indicator
14. Lean Manufacturing
15. Management Review
16. MBWA (Management by Walking Around)
17. MTBF (Mean Time Between Failures)
18. Operator Self-Control
19. Organisation
20. Organisational culture
21. PDCA (Plan–Do–Check–Act)
22. Process
23. Process capability
24. Process capability study
25. Process variation
26. PONC
27. Quality
28. Quality Costs
29. Quality management principles
30. Quality management processes
31. Quality Management System
32. Reliability
33. Risk
34. Six Sigma
35. Six Sigma project methodologies
36. Supplier Quality Assurance
37. Surveillance visits
38. Static and learning organisations
39. Systems thinking
40. Top management
41. Visual Management (VM)
42. World-Class Manufacturing

1. Best practice: A method or technique that has consistently shown effective results superior to those achieved with other means.

Best practice is a feature of accredited management standards such as ISO 9001 and the equivalent standard in China, GB/T 19001.

2. Blame culture: Where focus of blame for a fault habitually shifts to the person in an organisation, the organisation is referred to as having a blame culture. The result is that people within the organisation are unwilling to accept responsibility for mistakes or take risks due to a fear of criticism or prosecution. It is known to cause people to hide mistakes and to even drive people to be untruthful.

3. China Quality Award: The model for the "China Quality Award" follows the intent of renowned National Quality Awards such as the Deming Prize (Japan) and the European Quality Award and is based on the Malcolm Baldrige National Quality Award (USA).

The China Quality Award may be presented to organisations, projects and individuals.

Organisations are assessed in the following seven weighted point-based categories with 1000 points being 100%:

- Leadership (10%): role of senior leadership, organisation governance, and social responsibility
- Strategy (8%): strategy development and deployment
- Customer and market (9%): customer and market knowledge, customer relationships and satisfaction
- Resource (12%): human, financial, information and knowledge, technical, infrastructure, and relevant stakeholder relationship
- Process Management (11%): process identification, design and improvement
- Measurement, Analysis and Improvement (10%): measurement, analysis, review, improvement and innovation
- Results (40%): performance results and outcomes pertaining to product and service, customer and market, financial, resource, process effectiveness, and leadership

4. Continuous improvement and continual improvement: Ongoing efforts to improve products, services, processes, procedures and practices are both continuous and continual. Continuous improvement concerns gradual and ongoing changes, while continual improvement involves incremental step changes. When improvements occur in a big step change, this is referred to as a "breakthrough" improvement.

Improvement is not limited to quality initiatives; improvement in business strategy, business results and customer, employee and supplier relationships can be subject to continual improvement.

5. Critical-To-Quality (CTQ): A CTQ is a feature, characteristic or property of a part, assembly, sub-assembly, product, or process that has a direct and significant impact on its actual or perceived quality.

CTQs are the key measurable features, characteristics or properties of a product or process whose performance standards or specification limits must be met in order to meet customer and user needs.

6. Effectiveness: The extent to which the outputs of the processes meet the needs and expectations of its customers. Effectiveness is having the right output at the right place, at the right time, at the right price.

7. Efficiency: The extent to which the resources are minimized and waste is eliminated in the pursuit of effectiveness. Productivity is a measure of efficiency.

8. 5 Whys: This is a question-asking technique with the primary goal of determining the root cause of a defect or problem by repeating the question "Why?". Each question forms the basis of the next question.

Basing actions on symptoms is the worst possible practice, and by asking the question "Why" the symptoms can be separated from the causes of a problem. Using this technique can identify the root cause of a problem and subsequently lead to defining effective corrective actions. If it is found that multiple root causes are required to be uncovered, this question-asking technique must be repeated asking an appropriately different sequence of questions each time.

9. Function: A function is a specialised operation performed by a group of people, a department, a person or a machine. Design, sell, purchase, produce, are examples of functions, and, for instance the design function would be performed by a design department.

Often a number of functions located in departments are required to accomplish a process; for example, the process of "acquiring a bought-in product" may likely require the needed specialist knowledge and skills located in the purchasing, accounting, design, and supplier quality assurance departments.

10. ISO 9000: The ISO 9000 series or family of standards are a group of international consensus standards that address various aspects of quality management and comprise of ISO 9000 (terms and definitions), ISO 9001 (requirements) and ISO 9004 (continuous improvement). They are applicable to any organisation, regardless of size, type and activity.

Each ISO member country has their own entity authorized by ISO to manage the standards, and the standards used by each member country are all the same ISO 9000 quality documents and set of requirements.

In China, mandatory national standards are prefixed "GB" (Pinyin: *guó biāo*) and recommended standards are prefixed "GB/T" (Pinyin: *tuī jiàn*). The GB/T 19000 series is the family of Chinese national standards that are the direct equivalent of the ISO 9000 series.

- **ISO 9000:2015 (GB/T 19000:2016), Quality Management Systems – Fundamentals and Vocabulary:** Describes the fundamental concepts and principles of quality management and includes the terms and definitions that apply to all quality management and quality management system standards developed by ISO/TC 176.

- **ISO 9001:2015 (GB/T 19001:2016), Quality Management System – Requirements:** Sets out the requirements of a quality management system and is the only standard in the ISO 9000 family that organisations can be certified to. Using ISO 9001:2015 helps ensure that customers get consistent, good quality products and services, which in turn brings many business benefits.

Note that ISO 9002 and ISO 9003 were combined in the 2000 revision of ISO 9001 and are now obsolete.

- **ISO 9004:2009** [13] **(GB/T 19004:2011), Managing for the Sustained Success of an Organisation – A Quality Management Approach:** Focuses on how to make a quality management system more efficient and effective. It provides guidance to organisations to support the achievement of sustained success by a quality management approach. ISO 9004:2009 is not intended for certification, regulatory or contractual use.

11. ISO 19011:2011[13] (GB/T 19011:2013), Guidelines for Auditing Management Systems: In order to evaluate the continued effectiveness to their established and maintained ISO management systems, organisations are required to perform internal audits.

ISO 19011 explains the principles of auditing, the management of an audit programme and the performing of management system audits. It also provides guidance on the evaluation of competence of individuals involved in the audit process – this includes the person managing the audit programme, the auditors and audit teams.

12. KAIZEN [14]: KAIZEN concerns small improvements from the ongoing efforts of everybody. All levels of management, supervision and operator workforce share a responsibility for KAIZEN.

The principles behind KAIZEN are that the jobs in any organisation have the components of process control and process improvement. Process control involves the taking of action on deviations to maintain a given process state and is described in standard operating procedures (SOPs). On the other hand, process improvement requires modifying the process to produce better results. Successful verification of the improvement will result in the SOP being updated.

The fundamental idea behind KAIZEN is the PDCA Cycle:

---

[13] ISO release updated versions of ISO 9004 and ISO 19011 in 2018.
[14] KAIZEN is a registered trademark of the KAIZEN Institute, Ltd.

- **Plan:** Someone has an idea for doing the job better
- **Do:** Trials or experiments are done to investigate the idea
- **Check:** The results are evaluated to determine whether the idea produced the desired result
- **Act:** If the idea proved successful, the SOP will be changed

KAIZEN does not concern radical innovations but rather works to optimise and improve the existing system.

Generally KAIZEN is implemented via quality improvement teams at various levels in the organisation.

The Japanese approach is to integrate KAIZEN into management systems to ensure that it is done routinely.

13. Key Performance Indicator (KPI): Key Performance Indicators, also known as Key Success Indicators (KSI) are quantifiable measurements decided by management that reflect the critical success factors of an organisation. The KPIs are used to direct actions; measurements are based on legitimate data and are used to gauge performance in terms of meeting strategic and operational goals. KPIs vary between companies and industries, depending on their priorities or performance criteria.

14. Lean Manufacturing: Lean Manufacturing was developed primarily from the Toyota Production System (TPS); it is an important methodology of World-Class Manufacturing. Lean Manufacturing focuses on providing customer satisfaction in the most profitable way, with customer satisfaction taking the fore at all times. The idea is that everything that should be done must provide value to the customer and anything else is waste.

Although developed mainly within manufacturing, Lean can be applied to office based administrative functions or to service industries such as banking.

15. Management Review: The purpose of a management review is to evaluate the overall performance of an organisation's quality management system and to identify improvement opportunities.

Results of a management review should lead to actions that ensure the continued effectiveness and continuing improvement of the quality management system and key business processes.

In the past, many organisations held management reviews once per year, however at this frequency, needed changes and improvements were being recognised too late. It is now favoured to have quality management reviews done at regular intervals throughout the year. This enables the continual monitoring of processes and the timely intervention when outcomes are not as desired.

16. MBWA (Management by Walking Around): This is an informal and personal approach by managers and supervisors in work-related affairs of their subordinates, in contrast to distant management (via emails, notices, memos and formal meetings). In MBWA practice, managers spend some of their time making informal visits to the work areas, observing practices, listening to the employees, and encouraging a quality attitude and positive action. In doing this exercise the manager collects first-hand qualitative information, hears suggestions and complaints, and is able to keep a finger on the pulse of the organisation.

17. MTBF (Mean Time Between Failures): This is a basic measure of a component's, product's or system's reliability.

MTBF excludes downtime, i.e., the time spent waiting for repair, being repaired, being re-qualified, and other events that cause downtime such as inspections and preventive maintenance. MTBF measures only the time a component, product or system is available and operating and is typically expressed in units of hours; the higher the MTBF number, the higher the reliability of the component, product or system.

Note that where the component, product or system cannot be repaired during its mission, the measure is Mean Time to First Failure (MTFF).

18. Operator Self-Control: This is a status achieved when the operator can control the quality of his work. To achieve this status, the operator must be provided with the means needed to carry out his assigned job. The pre-requisites of the "Operator Self-Control" approach entails:

- having suitable operators selected and trained for the job;
- taking care that the operator can "make it right the first time" by providing the operator with appropriate information, instruction, work-area (space, environment, atmosphere), material, equipment, etc.;
- establishing a detection process so that the operator knows if he is going wrong;
- and providing the operator with suitable information, instruction and means to assess and measure quality conformance.

19. Organisation: A body of people such as a business, a company, an institution or a branch, division or department that is structured and managed to work together for a particular purpose or to meet a need.

20. Organisational culture: Organisational culture is also called corporate culture. This is determined by the shared assumptions, values, attitudes, beliefs, and customs that contribute to the social and psychological environment of an organisation. Organisational culture is unique to every organisation and influences communication, employee behaviour, and how employees perform their jobs.

21. PDCA, Plan–Do–Check–Act: Also known as the Shewhart Cycle, after its originator, Walter Shewhart, or the PDSA (Plan-Do-Study-Act) method. PDCA was popularized by W. Edwards Deming and therefore came to be known also as the Deming Cycle. PDCA is among the most widely used methods for continuous improvement. This is a four–step methodology for carrying out change. Just as a circle has no end, the PDCA Cycle should be repeated for continuous improvement.

- **Plan:** Identify an opportunity and plan for change.
- **Do:** Implement the change on a small scale.
- **Check:** Use data to analyse the results of the change and determine whether it made a difference.
- **Act:** If the change was successful, implement it and continuously assess your results. If the change did not work, begin the cycle again.

22. Process: A process is an activity or a unique combination of activities that takes an input, adds some form of value to it and provides measurable output to an internal or external customer. A process consumes resources, e.g., manpower, machines and energy, to convert inputs such as information and/or parts into outputs.

A process could be of a business or technical nature. Examples include developing products, receiving orders, purchasing supplies, welding a frame, heat treating a shaft, performing a non-destructive test.

23. Process capability: All processes have inherent or natural variation; if this inherent variation falls within the given limits of a specification or tolerance, the output of the process would have 100% conformance to the given specification or tolerance and the process would be deemed as capable of producing within given limits.

A Process Capability Index (Cp and Cpk) is the ratio of the process performance values to the specification limits. The prerequisite for determining Cp and Cpk is that the process must be stable (in statistical control) with performance values normally distributed within three standard deviations (3σ) on either side of its process average.

24. Process capability study: A process capability study concerns the measure of the variability of the output of a process, and comparison of that variability with a given specification or product tolerance.

25. Process variation: Dr Walter Andrew Shewhart identified that the sources of process variation can be grouped into "common" (natural or normal) and "special" (assignable) cause categories.

The "common" causes of variation are inherent in a process. They are almost always not controllable by process operators and are generally due to system or management-controllable issues.

The "special" causes of variation are process disturbance events. The operator can usually control or remove "special" causes of variation, provided the operator is made aware of them.

26. PONC: PONC is the acronym for "Price of Non-Conformance", and comprises those costs incurred before and after delivery or handover of products and/or services to the customer, that do not conform to customer requirements or needs.

PONC = Internal Failure Costs + External Failure Costs.

27. Quality: In technical usage, quality can have two meanings:
- The characteristics of a product or service that bear on its ability to satisfy stated or implied needs.
- A product or service free of deficiencies.

According to Joseph Juran[15], quality means "fitness for use"; according to Philip Crosby, it means "conformance to requirements." (Ref: ASQ)

There are eight exclusive dimensions of quality: conformance, performance, features, reliability, durability, serviceability, aesthetics, and perceived quality. These dimensions can be viewed separately, e.g., a product can have high performance but low reliability, and as working in conjunction with each other, e.g., a product can have durability and reliability.

28. Quality Costs: The cost arising from activities aimed at assuring quality and the cost of the loss incurred when quality is not achieved. There are four types of Quality Costs:
- **Prevention Costs**, the costs of all activities aimed at preventing, avoiding or minimizing poor quality in products and services.
- **Appraisal Costs**, the costs incurred before delivery or handover to customers associated with measuring, evaluating, or auditing products or services to assure conformance to quality standards and conformance requirements.
- **Internal Failure Costs**, the costs incurred before delivery or handover to customers resulting from products or services not conforming to requirements or customer needs.

---

[15] Dr Joseph M. Juran (December 24, 1904 – February 28, 2008), American engineer, author, lecturer and quality evangelist and management consultant.

- **External Failure Costs**, the costs incurred after delivery or handover to customers resulting from products or services not conforming to requirements or customer needs.

29. **Quality management principles:** ISO 9001:2015 builds on seven quality management principles; the rationale behind each of these is as follows:

(1) **Customer Focus.** Sustained success is achieved when an organisation attracts and retains the confidence of customers. Every aspect of customer interaction provides an opportunity to create more value for the customer. Understanding current and future needs of customers contributes to sustained success of the organisation.

(2) **Leadership.** Creation of unity of purpose and direction and engagement of people enable an organisation to align its strategies, policies, processes and resources to achieve its objectives.

(3) **Engagement of people.** To manage an organisation effectively and efficiently, it is important to involve all people at all levels and to respect them as individuals. Recognition, empowerment and enhancement of competence facilitate the engagement of people in achieving the organisation's quality objectives

(4) **Process approach.** The management system consists of interrelated processes. Understanding how results are produced by this system enables an organisation to optimise the system and its performance.

(5) **Improvement.** Improvement is essential for an organisation to maintain current levels of performance, to react to changes in its internal and external conditions and to create new opportunities.

(6) **Evidence-based decision making.** Decision making can be complex; it involves some uncertainty and often involves multiple types and sources of inputs as well as their interpretation. It is important to understand cause-and-effect relationships and potential unintended consequences. Facts,

evidence and data analysis lead to greater objectivity and confidence in decision making.

(7) **Relationship management.** Sustained success is more likely to be achieved when the organisation manages relationships with all of its interested parties to optimise their impact on its performance. Relationship management with its supplier and partner networks is of particular importance.

30. Quality management processes: These are the processes that are engaged in decisions about the implementation of the quality delivery process. "Quality management processes" are not processes exclusively carried out by quality department personnel, or processes exclusively performed by "management".

31. Quality Management System (QMS): A set of interrelated activities that direct and control an organisation in order to continually improve the effectiveness and efficiency of its performance. The focus is on achieving quality policy and objectives to meet customer requirements. These interrelated elements include programmes, plans, agreements, processes, methods, procedures, practices, tools, techniques, records and resources. An effective QMS enables an organisation to reduce and aim to eliminate non-conformance to specifications, standards and customer requirements in the most cost effective and efficient manner.

32. Reliability: The ability of an item to perform a required function under stated conditions for a stated period of time – the item could be a mechanical, electronic, or software product, system or component, a manufacturing process, or even a service. The term "reliability" is also used as a reliability characteristic denoting a probability of success or a success ratio. (ISO 8402:1986, Quality – Vocabulary)

33. Risk: A risk is an undesirable situation or circumstance that has two characteristics: it is both likely to occur and likely to cause negative consequences. The concept of risk is always future oriented: it worries about the impact that undesirable circumstances or situations could have in the future.

34. Six Sigma[16]: Six Sigma is a project-by-project driven methodology that follows a defined sequence of steps and has specific quantifiable value targets. The focus of Six Sigma is on statistically defect free products or processes and increased customer satisfaction, defect prevention, process cycle time reduction, and cost savings through waste reduction.

Six Sigma relies on proven methods, techniques and tools for process improvement; a small cadre of people are trained to a high level of proficiency in their application. These people become the in-house Six Sigma leaders and are known as Six Sigma Black Belts.

35. Six Sigma project methodologies: There are two Six Sigma project methodologies, both inspired by the PDCA methodology:

- **DMAIC (Define–Measure–Analyse–Improve–Control)**

  Used when a project's goal can be accomplished by improving an existing product, process or service.

  D: Define the goals of the new state (obtained from customers, shareholders and employees), the project business case, scope and due dates of deliverables.

  M: Measure the existing or "as-is" state; collect adequate data. Establish valid and reliable metrics to help monitor progress towards the defined goals.

  A: Analyse to identify ways to eliminate the gap between the "as-is" state and the "to-be" state. Determine the root cause of problems.

---

[16] Six Sigma is a Motorola registered trademark.
A sigma (σ) rating is a way of describing the maturity of a process by indicating the percentage of defect-free products it creates. The Six Sigma standard is 3.4 defects or problems per million opportunities*, however, organisations that apply the Six Sigma methodology tend to determine an appropriate sigma level for each targeted process or product and then strive to achieve this level.
*Note that at 6 σ, the area under the normal curve indicates 2 parts-per-billion, but 3.4 parts-per-million is actually the area indicated at 4.5 σ of the normal curve.

Use statistical tools to guide analysis. Identify resource requirements and major obstacles.

I: **Improve** or optimise to achieve the "to-be" state. Use project management tools to implement and statistical methods to validate the improvement.

C: **Control** the new state. Institutionalize by modifying, e.g., incentive schemes, policies, procedures, budgets. Use statistical methods to monitor the stability of the improvement.

- **DMADV (Define–Measure–Analyse–Design–Verify)**

  Used when the goal is the development of a new or radically re-designed product, process or service.

  D: **Define** the goals of the design activity that are consistent with customer demands and the enterprise strategy.

  M: **Measure** / identify critical-to-quality characteristics, product capabilities, production process capability, and risks. Translate into goals.

  A: **Analyse** the options available for meeting the goals; include the performance of similar best-in-class designs.

  D: **Design** the new product, service or process. Use for example, predictive models and prototypes to validate the design's effectiveness in meeting goals.

  V: **Verify** the designs effectiveness in the real world.

Some practitioners have combined ideas from Six Sigma and Lean Manufacturing to create a **Lean Six Sigma** methodology aimed at promoting business and operational excellence.

36. Supplier Quality Assurance: This aims to provide confidence in a supplier's or sub-contractor's ability to deliver products or services that will satisfy the purchaser's needs. Supplier Quality Assurance is achievable through a relationship between the purchaser and the supplier that is established through supplier quality management practices.

37. Surveillance visits: These are quality control-orientated visits to the supplier's factory to ensure that the supplier's personnel are following what has been specified and agreed in contract documents.

38. Static and learning organisations: Some organisations are labelled as static organisations and others as learning organisations.

- **Static organisations** have fixed practices and do not change significantly over time. Typically they have a strict chain of command with senior management doing the thinking for the organisation; communication is predominantly one-way downward; personnel are managed through coercive power and discouraged from thinking, and become reserved. Static organisations have difficulty remaining competitive in the modern global economy.

- **Learning organisations** facilitate the learning of its members and are continuously developing, adapting, and transforming themselves in response to the challenges that they face to remain relevant and competitive in the modern global business environment. They encourage employees to generate positive and constructive ideas, and challenge all employees to tap into their inner potential.

39. Systems thinking: All systems are made up of component parts. In organisations the system components are people, structures and processes. "Systems thinking" is based on the understanding that the component parts of a system can best be understood in the context of their relationships with each other and with other systems, rather than in isolation. "Systems thinking" is sometimes referred to as the "cornerstone" of the learning organisation. It is an approach used in problem solving.

The concepts of the learning organisation and systems thinking are rooted in the work conducted by Peter Senge[17].

---

[17] Dr Peter Michael Senge (born 1947) systems scientist, author, lecturer at MIT Sloan School of Management and founder of the Society for Organisational Learning.

40. Top management: Persons who direct an organisation at its highest level with primary responsibility for decisions and activities affecting the organisation in the long term. These persons are typically company officers and director-level executives.

41. Visual Management (VM): Visual Management comprises of visual displays for conveying information to people in an area and visual controls to control or guide actions. VM helps with communicating and sharing information such that people understand situations at a glance, and makes problems, abnormalities, or deviations from standards visible to everyone.

42. World-Class Manufacturing: World-Class Manufacturing is a collection of best practice methodologies and techniques. The main parameters that determine World-Class Manufacturers are quality, operational efficiency, waste reduction, cost effectiveness, flexibility and innovation. World-Class Manufacturers employ methodologies and techniques as;

- Continuous improvement, root cause analysis and PDCA
- employee involvement
- Operator Self-Control
- Visual Management
- Statistical Process Control
- Standardisation
- Key Performance Indicators, and skills and competency reviews
- 5S/5C
- Overall Equipment Effectiveness
- Lean Manufacturing
- Cellular Manufacturing
- Total Productive Maintenance
- rapid changeover of tooling (SMED, Single-Minute Exchange of Die)
- value-stream mapping

APPENDIX 2: LIST OF TABLES

APPENDIX 3: LIST OF FIGURES

INDEX